MW01154811

Variational Methods In Mechanics

Variational Methods in Mechanics

Toshio Mura

Tatsuhito Koya

Department of Civil Engineering
and Applied Mathematics
Northwestern University

New York *Oxford*
Oxford University Press
1992

Oxford University Press

Oxford New York Toronto
Delhi Bombay Calcutta Madras Karachi
Kuala Lumpur Singapore Hong Kong Tokyo
Nairobi Dar es Salaam Cape Town
Melbourne Auckland

and associated companies in
Berlin Ibadan

Published by Oxford University, Inc.
200 Madison Avenue, New York, New York 10016

Library of Congress Cataloguing-in-Publication Data
Mura, Toshio, 1925–
Variational methods in mechanics / Toshio Mura, Tatsuhito Koya.
p. cm. Includes bibliographical references and index.
ISBN 0-19-506830-0
1. Finite element method.
2. Mechanics, Applied—Mathematics.
3. Variational principles.
I. Koya, Tatsuhito. II. Title.
TA347.F5M87 1992. 620.1–dc20
91-18834

1 3 5 7 9 8 6 4 2

Printed in the United States of America
on acid-free paper

To Courtney and Stephanie

Preface

This book stems from a course on variational principles in mechanics that Professor T. Mura has been teaching for about 25 years at Northwestern University. Because he has been involved in the Material Research Center, half of the students in the course have been materials science students. Professor Mura has paid special attention to teaching these students who are often inexperienced in mechanics and mathematics.

When T. Mura was a graduate student at the University of Tokyo, his main accomplishment, besides writing his thesis, was reading R. Courant's and D. Hilbert's *Methoden der Mathematischen Physik*. This book, therefore, is very much influenced by that masterpiece. It is the intention of the authors to explain here the essence of Courant and Hilbert in practical, commonsense terms.

This book has been written primarily for students in engineering schools and for practicing mechanical engineers. Most of the differential equations appearing in science and engineering can be derived from minimum, maximum, and stationary conditions for some integrals. Integrands of the integrals contain unknown functions appearing in the differential equations. These extremums are taken with respect to the unknown functions. This extremum principle is a unique aspect of nature and also provides a powerful approximation or numerical method for solving differential equations. The finite element method is based upon this principle.

The variational methods concern calculations associated with this extremum principle, and this book emphasizes applications of the methods in science and engineering and minimizes mathematical arguments about the existence theorems, continuity conditions, convergency discussions, and so on. Mechanics is a general topic of the applications in this book.

The authors are not sure whether the title of this book should be *Variational Methods* or *Variational Principles*. An alternative title could be *Calculus of Variations*. In any case, the authors have emphasized applications to problems in mechanics. Mechanics itself could be regarded as nothing more than a variational principle.

The variational principles are attractive primarily for their beauty of form and spirit. Nature follows a principle in which certain energies try to be minimal. The variational methods are also convenient because they provide powerful approximation methods. The recent advance of the finite element method is based upon these variational methods.

Due to the recent development of numerical analysis by computers, the manuscript originally written by Mura, who has never used computers, needed proper modifications. Coauthor T. Koya took care of these modifications.

During the last three decades, research on composite materials has become a very popular branch of mechanics. Fortunately, Professor John Willis of the University of Bath in England, who is a leader in this field, agreed to contribute Chapter 21 on the application of the variational methods to the evaluation of overall properties of composite materials.

Evanston, Ill. T. M.
December 1991 T. K.

Acknowledgments

We wish to express our thanks to Dr. Y. Nakasone for reading the manuscript and to Ms. Marybeth Nugent and Ms. Mary Hill for their skillful typing and patience.

We thank Dr. Michio Inokuchi of Argonne National Laboratory who has taken over Mura's classes whenever it was necessary. Inokuchi has shown the simple but essential spirit behind the variational principles applied to physics. In particular, his contributions in Chapter 18, the General Use of Lagrange Multipliers, are very invigorating.

We express our special acknowledgment to Professor John Willis of the University of Bath in England who kindly supplied Chapter 21, "Bounds for the Overall Properties of Anisotropic Composites."

We also extend thanks to all the students who took our course, the Variational Principles of Mechanics. Their suggestions and comments were valuable and helped to refine many sections in this book.

Finally, we wish to thank the generous attitude and cooperation of the publisher.

Contents

Variational Methods In Mechanics

1

Maxima and Minima of a Function

The calculus of variation (the variational method) is a natural extension of the maximum and minimum theory of ordinary functions.

Consider a continuous function $f(x)$ defined in a closed domain G and discuss a point x^*, where $f(x)$ takes a local maximum (see Fig.1.1a), a local **minimum** (Fig.1.1b), or an **inflection** (Fig.1.1c). The maximum or minimum value of the function is called the **extremum**. The function is called **stationary** at x^* in Fig. 1.1a, b, and c. The necessary condition for $f(x)$ to be stationary is

$$f'(x) = 0. \tag{1.1}$$

The necessary and sufficient conditions are obtained by adding the following conditions to (1.1): $f''(x) > 0$ for minimum, $f''(x) < 0$ for maximum, and $f''(x) = 0$ for inflection. When $f(x,y,z,\ldots)$ is a function of many variables $x,y,z,\ldots$, the necessary conditions for f to be stationary are

$$f_x = 0, \quad f_y = 0, \quad f_z = 0, \ldots. \tag{1.2}$$

Otherwise stated, we assume in this book the maximum or minimum always exists. The maximum, for instance, appears at $x = x^*$ in Fig. 1.1a.

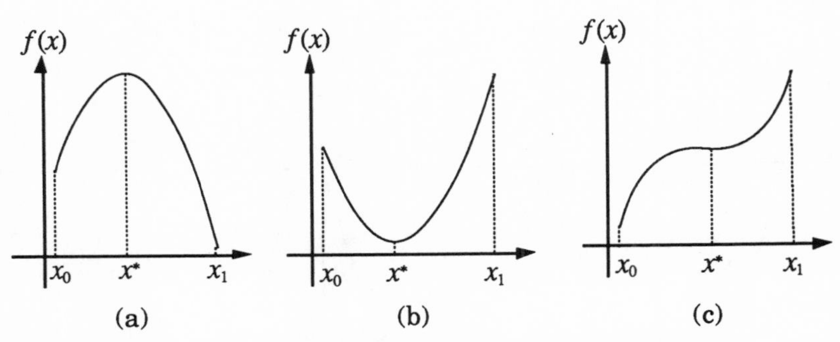

Figure 1.1. Function $f(x)$ is stationary at point x^*. (a) Maximum, (b) minimum, (c) inflection.

Example 1.1

Find the path of light, starting from a given point $P(x_1,y_1)$ and arriving at point,$Q(x_0,y_0)$ where medium 1 and medium 2 are different homogeneous media in which the velocities of light are c_1 and c_2, respectively (Fig. 1.2).

Acoording to the **principle of Fermat,** the time elapsed in the passage of light between two fixed points is a minimum with respect to possible paths connecting two points. The time T is

$$T=\frac{\sqrt{(x_1-x)^2+y_1^2}}{c_1}+\frac{\sqrt{(x-x_0)^2+y_0^2}}{c_2} \tag{1.3}$$

The equation $dT/dx=0$ leads to

$$\frac{(x_1-x)}{c_1\sqrt{(x_1-x)^2+y_1^2}}=\frac{(x-x_0)}{c_2\sqrt{(x-x_0)^2+y_0^2}} \tag{1.4}$$

or

$$\frac{\sin\theta_1}{c_1}=\frac{\sin\theta_2}{c_2}. \tag{1.5}$$

The condition (1.5) is called **Snell's law.**

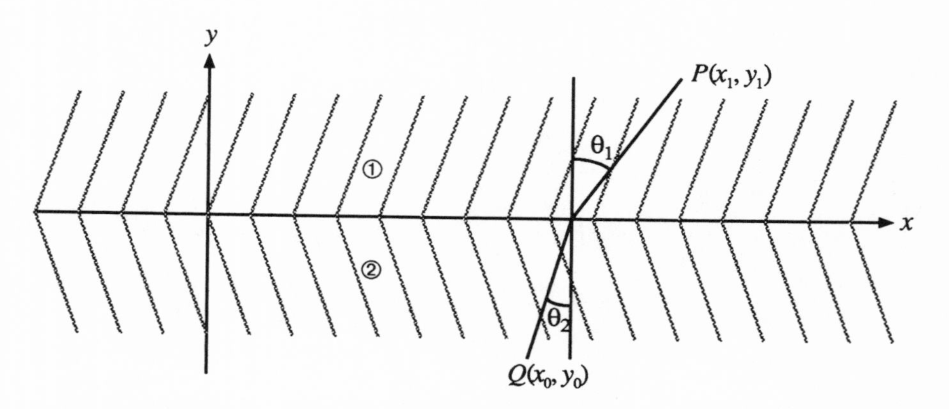

Figure 1.2. Fermat's principle. In media 1 and 2, light travels at different velocities.

Lagrange Multiplier

Consider the extremum problem of $f(x,y,z)$ under the **subsidiary condition**

$$g(x,y,z)=0. \tag{1.6}$$

The necessary condition for $f(x,y,z)$ to be stationary is

$$df = f_x dx + f_y dy + f_z dz = 0. \tag{1.7}$$

Since the variables x, y, and z are not independent, the coefficients of dx, dy, and dz in (1.7) are not zero. An elementary way of solving the problem is to express z in terms of x and y by solving (1.6) and to substitute for z in $f(x,y,z)$. Then $f(x,y,z(x,y))$ becomes a function of two independent variables x and y. Condition (1.7) thus can be written as

$$\left(f_x + f_z \frac{\partial z}{\partial x}\right)dx + \left(f_y + f_z \frac{\partial z}{\partial y}\right)dy = 0. \tag{1.8}$$

Finally, the extremum condition is obtained as

$$f_x + f_z \frac{\partial z}{\partial x} = 0, \quad f_y + f_z \frac{\partial z}{\partial y} = 0. \tag{1.9}$$

This elementary method, however, is inconvenient because, in general, Eq. (1.6) cannot be solved analytically for z. The most convenient method is the **Lagrange method**.

From Eq. (1.6) we have

$$dg = g_x dx + g_y dy + g_z dz = 0. \tag{1.10}$$

Introducing a parameter λ, the Lagrange multiplier, we can write

$$df + \lambda dg = (f_x + \lambda g_x)dx + (f_y + \lambda g_y)dy + (f_z + \lambda g_z)dz = 0. \tag{1.11}$$

The parameter λ is chosen so that

$$f_z + \lambda g_z = 0. \tag{1.12}$$

Then, the remaining expression becomes

$$(f_x + \lambda g_x)dx + (f_y + \lambda g_y)dy = 0. \tag{1.13}$$

Since dx and dy can be independent, their coefficients are zero,

$$f_x + \lambda g_x = 0, \quad f_y + \lambda g_y = 0. \tag{1.14}$$

The four equations (1.6), (1.12), and (1.14) are used to solve the extremum point x, y, z and the Lagrange multiplier λ.

Generally, when $f(x_1, x_2, \ldots, x_n)$ is to be stationary under the subsidiary conditions ($m < n$),

$$g_1(x_1, x_2, \ldots, x_n) = 0, \ldots, g_m(x_1, x_2, \ldots, x_n) = 0, \tag{1.15}$$

$$\delta F = F_{x_1}\delta x_1 + F_{x_2}\delta x_2 + F_{x_3}\delta x_3 + \cdots + F_{x_n}\delta x_n = 0 \tag{1.16.1}$$

or

$$\begin{aligned}&\left(f_{x_1} + \sum_{i=1}^{m}\lambda_i g_{i,x_1}\right)\delta x_1 + \left(f_{x_2} + \sum_{i=1}^{m}\lambda_i g_{i,x_2}\right)\delta x_2 \\ &+ \left(f_{x_{n-m}} + \sum_{i=1}^{m}\lambda_i g_{i,x_{n-m}}\right)\delta x_{n-m} + \cdots + \left(f_{x_1} + \sum_{i=1}^{m}\lambda_i g_{i,x_n}\right)\delta x_n = 0.\end{aligned} \tag{1.16.2}$$

$\lambda_1, \lambda_2, \lambda_3, \ldots, \lambda_m$ are determined from

$$\begin{aligned}f_{x_{n-m}} + \sum_m \lambda_m g_{m,x_{n-m}} &= 0 \\ f_{x_{n-m+1}} + \sum_m \lambda_m g_{m,x_{n-m+1}} &= 0 \\ f_{x_{n-m+2}} + \sum_m \lambda_m g_{m,x_{n-m+2}} &= 0 \\ f_{x_{n-m+3}} + \sum_m \lambda_m g_{m,x_{n-m+3}} &= 0 \\ f_{x_n} + \sum_m \lambda_m g_{m,x_n} &= 0\end{aligned} \tag{1.16.3}$$

Then (1.16.2) becomes

$$\left(f_{x_1}+\sum_m \lambda_m g_{m,x_1}\right)\delta x_1+\left(f_{x_2}+\sum_m \lambda_m g_{m,x_2}\right)\delta x_2 + \cdots + \left(f_{x_2}+\sum_m \lambda_m g_{m,x_{n-m}}\right)\delta x_{n-m}=0 \tag{1.16.4}$$

Since $\delta x_1, \delta x_2, \ldots, \delta x_{n-m}$ are independent, we have

$$\begin{aligned} f_{x_1}+\sum_m \lambda_m g_{m,x_1}&=0\\ f_{x_2}+\sum_m \lambda_m g_{m,x_2}&=0\\ f_{x_3}+\sum_m \lambda_m g_{m,x_3}&=0\\ &\cdots\\ f_{x_{n-m}}+\sum_m \lambda_m g_{m,x_{n-m}}&=0. \end{aligned} \tag{1.16.5}$$

Symbol δ in this context indicates an infinitesimal change. Finally, we can write (1.16.3) and (1.16.4) as

$$F_{x_1}=0,\quad F_{x_2}=0,\quad \ldots,\quad F_{x_n}=0 \tag{1.17}$$

and

$$F_{\lambda_1}=0,\quad F_{\lambda_2}=0,\quad \ldots,\quad F_{\lambda_m}=0. \tag{1.18}$$

The conditions (1.18) are identical to (1.15).

Example 1.2

Find the maximum volume of a box which has no cover (Fig. 1.3). The surface area of the box is given. The volume is expressed as $V(x,y,z) = xyz$ under the condition that the area $A(x,y,z) = 2xz + 2yz + xy =$ constant.

$$F = f + \lambda g = xyz + \lambda(2xz + 2yz + xy - C). \tag{1.19}$$

The simultaneous equations (1.17) and (1.18) become

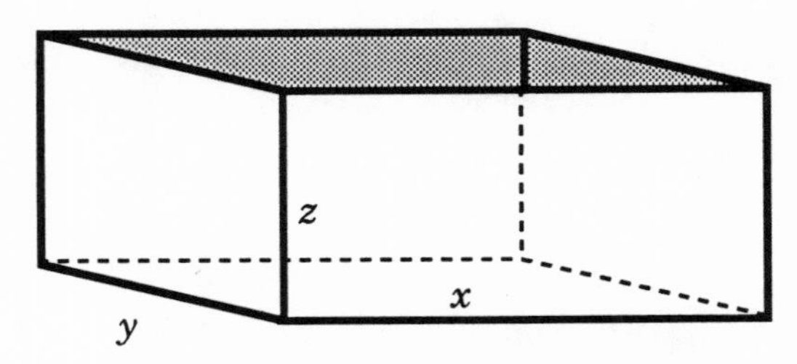

Figure 1.3. A box without cover.

$$\begin{aligned}\frac{\partial F}{\partial x} &= yz + \lambda(2z + y) = 0,\\ \frac{\partial F}{\partial y} &= xz + \lambda(2z + x) = 0,\\ \frac{\partial F}{\partial z} &= xy + \lambda(2x + 2y) = 0,\end{aligned} \tag{1.20}$$

The solution of (1.20) is

$$x = y = 2z = \sqrt{\frac{C}{3}}, \quad \lambda = -\frac{\sqrt{C}}{4\sqrt{3}}. \tag{1.21}$$

The corresponding maximum of V is

$$V = \frac{C^{\frac{3}{2}}}{6\sqrt{3}}. \tag{1.22}$$

It is rather complicated to prove analytically that this extremum is the maximum. However, we can easily see that it is the maximum when a special value of C is taken as 3 and values of V in the neighborhood of $x = y = 2z = 1$, satisfying condition $A(x,y,z) = C$, are compared with $V = ½$.

Example 1.3

Find the number n_i of atoms in energy level E_i that minimizes the free energy

$$G = E - TS, \tag{1.23}$$

where

$$E = n_1E_1 + n_2E_2 + \cdots \qquad : \text{internal energy}$$

$$S = k \log \frac{n!}{n_1!\ n_2!\ n_3!\ \cdots} \qquad : \text{entropy}$$

$$n = n_1 + n_2 + \cdots \qquad : \text{number of atoms}$$

E_i is the energy level of n_i number of atoms, T is the absolute temperature, and k is the Boltzmann constant. T, E_i , k, and n are given constants. n and n_i are large enough to assume log n_i ! = n_i (log n_i − 1). We have

$$F = n_1E_1 + n_2E_2 + \cdots - Tk\{n(\log n - 1) - n_1(\log n_1 - 1) \\ - n_2(\log n_2 - 1) - \cdots\} + \lambda(n - n_1 - n_2 - \cdots),$$

$$\frac{\partial F}{\partial n_i} = E_i + kT \log n_i - \lambda = 0, \tag{1.24}$$

$$n_i = \exp\left(\frac{\lambda - E_i}{kT}\right),$$

$$n = \exp\left(\frac{\lambda}{kT}\right)\sum_i \exp\left(-\frac{E_i}{kT}\right)$$

and finally

$$n_i = n\left(\frac{\exp\left(-\frac{E_i}{kT}\right)}{\sum_j \exp\left(-\frac{E_j}{kT}\right)}\right). \tag{1.25}$$

Problems

1.1. Find the n-gon which has the maximum area under the condition that every corner of the n-gon is on a common circle.

1.2. Find the maximum and minimum distances between an arbitrary point and a given sphere.

1.3. The ground state of energy of an electron with mass m confined in a box with sides a, b, and c is expressed as

$$E = \frac{h^2}{8m}\left(\frac{1}{a^2} + \frac{1}{b^2} + \frac{1}{c^2}\right).$$

When the volume of the box is kept constant, find the sides of the box to give a minimum value of E.

1.4. Find the shortest distance from the origin of coordinates to the hyperbola,

$$x^2 + 8xy + 7y^2 = 225.$$

1.5. Find the maximum value of the square of the vector product

$$(\mathbf{A} \times \mathbf{B})^2,$$

where

$$\|\mathbf{A}\| + \|\mathbf{B}\| = k, \quad \mathbf{C} \cdot \mathbf{A} = 0, \quad \mathbf{C} \cdot \mathbf{B} = 0.$$

1.6. Apply the principle of Fermat to the sound propagation between points P and Q in Fig. 1.2, where the elastic media (1) and (2) have different elastic moduli. Find two locations of R.

1.7. Prove that the system of equations (1.6), (1.12), and (1.14) is equivalent to (1.9).

1.8. Prove analytically that the problem in Example 1.2 is a maximum problem.

1.9. Prove that problem in Example1.3 is a minimum problem.

2

The Euler Equation I

We now deal with the problem of finding $y(x)$ that makes the following integral take a minimum value

$$I[y]=\int_{x_0}^{x_1}\left[p(x)(y')^2+q(x)y^2+2f(x)y\right]dx \tag{2.1}$$

with the prescribed boundary conditions,

$$y(x_0)=y_0, \quad y(x_1)=y_1, \tag{2.2}$$

where p, q, and f are given continuous functions in $x_0 \le x \le x_1$, and $p > 0$, $q > 0$. The conditions $p > 0$ and $q > 0$ are necessary for $I[y]$ to be a minimum problem, as will be seen later in this section. Values of the integral (2.1) depend on the function $y(x)$. Therefore, $I[y]$ is a function of $y(x)$. $I[y]$ is called the **functional** with **argument** y.

Denote the solution of the problem by $y = y(x)$ and a **comparison function** (or **trial function**) by

$$y=y(x)+\alpha\eta(x), \tag{2.3}$$

where α is a parameter and $\eta(x)$ is an arbitrary continuous function that vanishes at the boundaries,

$$\eta(x_0)=0, \quad \eta(x_1)=0. \tag{2.4}$$

When $\alpha = 0$, the comparison function becomes the solution. Since any comparison function must be identical to the solution at the boundaries, the last condition (2.4) is required for $\eta(x)$. As seen in Fig. 2.1, any **admissible comparison function** that gives a value of I in the neighborhood of the solution can be expressed in the form shown by (2.3).

When (2.3) is substituted into (2.1), we have

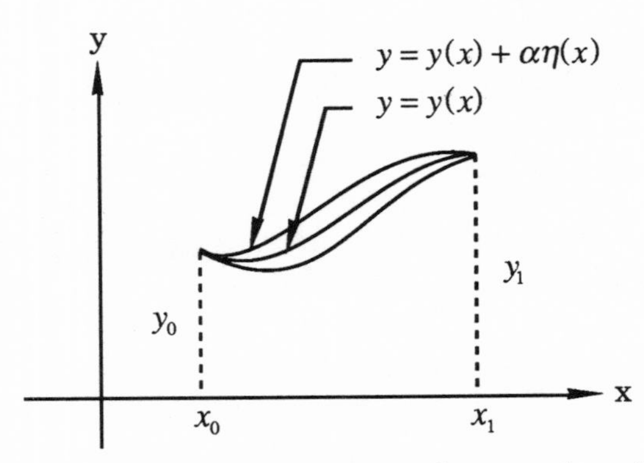

Figure 2.1. Stationary function $y = y(x)$ and comparison function $y = y(x) + \alpha\eta(x)$.

$$\begin{aligned}
I[y(x)+\alpha\eta(x)] &= \int_{x_0}^{x_1}\{p(x)[y'(x)+\alpha\eta'(x)]^2 + q(x)[y(x)+\alpha\eta(x)]^2 \\
&\qquad +2f(x)[y(x)+\alpha\eta(x)]\}\,dx \\
&= \int_{x_0}^{x_1}\{p(x)[y'(x)]^2 + q(x)y^2(x) + 2f(x)y(x)\}\,dx \\
&\qquad +2\alpha\int_{x_0}^{x_1}[p(x)y'(x)\eta'(x) + q(x)y(x)\eta(x) \\
&\qquad +f(x)\eta(x)]\,dx \\
&\qquad +\alpha^2\int_{x_0}^{x_1}\{p(x)[\eta'(x)]^2 + q(x)\eta^2(x)\}\,dx
\end{aligned} \tag{2.5}$$

or

$$I[y(x)+\alpha\eta(x)] - I[y(x)] = \delta I + \delta^2 I, \tag{2.6}$$

where

$$\delta I = 2\alpha\int_{x_0}^{x_1}[p(x)y'(x)\eta'(x) + q(x)y(x)\eta(x) + f(x)\eta(x)]\,dx \tag{2.7}$$

and

$$\delta^2 I = \alpha^2\int_{x_0}^{x_1}\{p(x)[\eta'(x)]^2 + q(x)\eta^2(x)\}\,dx \tag{2.8}$$

δI and $\delta^2 I$ are called the **first variation** and **second variation**, respectively. Since α is arbitrary, the necessary condition for $I[y(x)]$ to be a minimum is

$$\delta I = 0. \tag{2.9}$$

The first term in the integrand in (2.7) is integrated by parts,

$$\begin{aligned}\int_{x_0}^{x_1} py'\eta'\,dx &= [py'\eta]_{x_0}^{x_1} - \int_{x_0}^{x_1} (py')'\eta\,dx \\ &= -\int_{x_0}^{x_1} (py')'\eta\,dx.\end{aligned} \tag{2.10}$$

The boundary values in (2.10) vanish because of (2.4). Then the condition (2.9) is written as

$$\int_{x_0}^{x_1}\left\{-[p(x)y'(x)]' + q(x)y(x) + f(x)\right\}\eta(x)\,dx = 0. \tag{2.11}$$

Since (2.11) must hold for any $\eta(x)$, the coefficient of $\eta(x)$ must vanish,

$$-[p(x)y'(x)]' + q(x)y(x) + f(x) = 0. \tag{2.12}$$

The solution of the problem may be obtained by solving the differential equation (2.12) with the boundary conditions (2.2). The above equation is called the **Euler (Euler-Lagrange) equation**.

The Euler equation corresponds to $f'(x)=0$ in (1.1) for an ordinary function $f(x)$. The condition (2.12) is a necessary condition for $I[y]$ to be minimum or maximum.

The second variation $\delta^2 I$ is positive as long as $p > 0$ and $q > 0$, and the problem is guaranteed as a minimum problem. This corresponds to $f''(x) > 0$ for a minimum problem for an ordinary function $f(x)$.

Example 2.1

Find a function $y(x)$ that minimizes the functional

$$I[y] = \int_0^{\pi}\left[(y')^2 + y^2 + 2xy\right]dx \tag{2.13}$$

and satisfies the conditions

$$y(0) = 1, \quad y(\pi) = 0. \tag{2.14}$$

By satisfying this functional to that of (2.1), we see that

$$p(x)=1, \quad q(x)=1, \quad f(x)=x.$$

Therefore, (2.12) gives the Euler equation

$$y''-y-x=0.$$

The solution of this differential equation is

$$y=c_1\cosh x+c_2\sinh x-x.$$

From the given boundary conditions, it follows that

$$c_1=1, \qquad c_2=\frac{\pi-\cosh\pi}{\sinh\pi},$$

and hence, the solution is

$$y=\cosh x+\frac{\pi-\cosh\pi}{\sinh\pi}\sinh x-x. \tag{2.15}$$

The following **fundamental lemma** of the calculus of variations justifies the derivation of (2.12) from (2.11). If the relation $\int_{x_0}^{x_1}\rho(x)\eta(x)\,dx=0$, with $\rho(x)$ that is a **piecewise continuous function** of x, holds for all continuous functions $\eta(x)$ that vanish on the boundary, it follows that $\rho(x)=0$ identically. Let us suppose that $\rho(x)$ is different from zero in, say $G=\xi_0<x<\xi_1$, in which $\rho(x)$ is positive. We take the function $\eta(x)=(x-\xi_0)^4(x-\xi_1)^4$ in G and $\eta(x)=0$ outside this interval. Then we have $\int_{x_0}^{x_1}\eta\rho\,dx>0$ in contradiction to the hypothesis.

A function is piecewise continuous in a domain G when G may be subdivided into a finite number of domains such that, in the interior of each domain, the function is continuous and approaches a finite limit as a point on the boundary of these domains is approached from its interior. The function y shown in Fig. 2.2 is piecewise continuous but not y' since $y'=\infty$ at $x=0$.

According to the fundamental lemma, $\{-(py')'+qy+f\}$ must be a piecewise continuous function of x in order to conclude (2.12). This restriction will be relaxed by the **theorem of Haar** explained in Chapter 7.

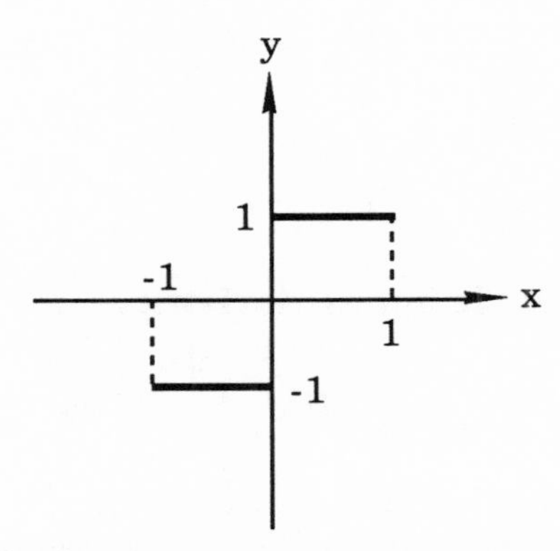

Figure 2.2. A piecewise continuous function.

Variational Derivatives

$\alpha\eta(x)$ and $\alpha\eta'(x)$ in (2.7) are written as δy and $\delta y'$, respectively. δy is a **variation** of y. Then (2.7) is written as

$$\delta I = 2\int_{x_0}^{x_1} (py'\delta y' + qy\delta y + f\delta y)\,dx. \tag{2.16}$$

Generally, when the functional to be extremized is given by

$$I[y] = \int_{x_0}^{x_1} F(x, y, y')\,dx, \tag{2.17}$$

where F is a given function of x, y, y', and is twice continuously differentiable with respect to its three arguments, the first variation of I becomes

$$\delta I = \int_{x_0}^{x_1} \left(F_y \delta y + F_{y'} \delta y'\right) dx. \tag{2.18}$$

When the boundary conditions $y(x_1) = y_1$ and $y(x_0) = y_0$ are given, $[\delta y]_{x_0}^{x_1} = 0$. The stationary condition becomes

$$\delta I = \int_{x_0}^{x_1} \left[F_y - \left(F_{y'}\right)' \right] \delta y\,dx = 0. \tag{2.19}$$

The Euler equation is obtained as

$$[F]_y = F_y - (F_{y'})' = 0 \tag{2.20}$$

or

$$F_y - F_{y'x} - F_{y'y}y' - F_{y'y'}y'' = 0. \tag{2.21}$$

$[F]_y$ is known as the **variational derivative** of F with respect to y. The continuity conditions given to F with respect to the arguments were necessary for the derivation of (2.21) from (2.19) by the fundamental lemma.

The second variation is obtained by the Taylor expansion of $I[y + \delta y]$ about $I[y]$. The Taylor expansion of function $F(x, y+\delta y, y'+\delta y')$ about $F(x,y,y')$ is

$$\begin{aligned} F(x,y+\delta y,y'+\delta y') &= F(x,y,y') + F_y(x,y,y')\delta y + F_{y'}(x,y,y')\delta y' \\ &+ \tfrac{1}{2}F_{yy}(x,y,y')(\delta y)^2 + \tfrac{1}{2}F_{y'y'}(x,y,y')(\delta y')^2 \\ &+ F_{yy'}(x,y,y')\,\delta y\,\delta y' \end{aligned} \tag{2.22}$$

where higher order terms are neglected. Then,

$$\begin{aligned} I[y+\delta y] &= I[y] + \int_{x_0}^{x_1}(F_y\delta y + F_{y'}\delta y')dx \\ &+ \tfrac{1}{2}\int_{x_0}^{x_1}\left[F_{yy}(\delta y)^2 + 2F_{yy'}\delta y\ \delta y' + F_{y'y'}(\delta y')^2\right]dx \end{aligned} \tag{2.23}$$

Then we have

$$I[y+\delta y] - I[y] = \delta I + \delta^2 I, \tag{2.24}$$

where

$$\delta^2 I = \tfrac{1}{2}\int_{x_0}^{x_1}\left[F_{yy}(\delta y)^2 + 2F_{yy'}\delta y\delta y' + F_{y'y'}(\delta y')^2\right]dx. \tag{2.25}$$

The extremum problem of $I[y]$ is a minimum problem if $\delta^2 I > 0$, and a maximum problem if $\delta^2 I < 0$. One of the necessary conditions for $\delta^2 I > 0$ is

$$F_{y'y'} > 0, \tag{2.26}$$

due to the following reason. Since δy is arbitrary, we chose such δy

as shown in Fig. 2.3, namely

$$\delta y = \begin{cases} \sqrt{\sigma}\left(1+\dfrac{x-\beta}{\sigma}\right) & \text{for } \beta - \sigma \le x \le \beta \\ \sqrt{\sigma}\left(1-\dfrac{x-\beta}{\sigma}\right) & \text{for } \beta \le x \le \beta + \sigma \\ 0 & \text{everywhere else.} \end{cases} \tag{2.27}$$

If we now let σ tend to zero, the first two terms of the integral (2.25) tend to zero, while the limit of the third term is the value of $2F_{y'y'}$ at $x = \beta$. Thus this value must be positive as shown in (2.26) when $\delta^2 I > 0$. The necessary condition (2.26) is called the **Legendre condition**.

The differential equation (2.20) is rewritten as

$$\frac{\left(F - F_{y'}y'\right)' - F_x}{y'} = 0. \tag{2.28}$$

Thus, if $F(x, y, y')$ does not explicitly depend on x ($F_x = 0$), we have

$$F - F_{y'}y' = \text{constant } c \tag{2.29}$$

from which y' is determined as a function of y and c.

If $F = (x, y)$, (2.28) becomes simply

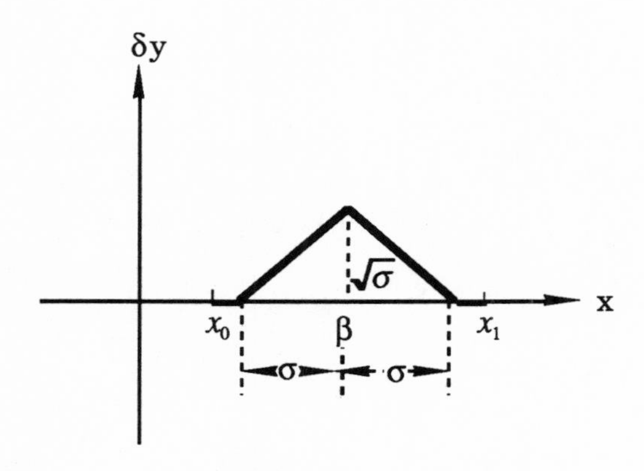

Figure 2.3. A choice of δy.

$$F_y = 0. \tag{2.30}$$

If $F = F(x, y')$, (2.28) becomes

$$\left(F_{y'}\right)' = 0 \tag{2.31}$$

and, therefore,

$$F_{y'} = \text{constant}. \tag{2.32}$$

Example 2.2

According to Fermat's principle, the path of a light ray in an inhomogeneous two dimensional medium in which the velocity of light $v(y)$ is determined by the minimum problem of

$$I[y] = \int_0^{\ell} \frac{\sqrt{1+y'^2}}{v(y)}\, dx \tag{2.33}$$

with boundary conditions $y(0) = 0$ and $y(\ell) = h$. Equation (2.29) leads to

$$\frac{1}{v\sqrt{1+y'^2}} = c. \tag{2.34}$$

When v is given as an explicit function of y, (2.34) yields $y' = y'(y)$ and, therefore, $x = \int_0^y dy/y'(y)$.

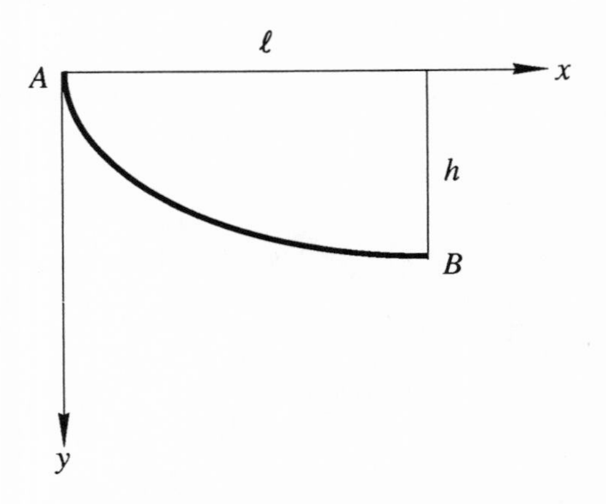

Figure 2.4. Brachistochrone, catenary.

Closely related to the problem of the light is the **brachistochrone problem**. Two points, $A(0, 0)$ and $B(\ell, h)$ with $h > 0$, are to be connected by a curve along which a frictionless mass point moves in the shortest possible time from A to B under gravity acting in the y direction (see Fig. 2.4). The initial speed of the mass point is zero. After falling a distance y, the point has the speed $\sqrt{2gy}$, according to elementary mechanics, where g is the magnitude of acceleration due to gravity. Then, the solution of (2.29) is obtained as

$$x = a(\theta - \sin\theta), \quad y = a(1 - \cos\theta), \quad 0 \le \theta \le \theta_s, \tag{2.35}$$

where $h = a(1 - \cos\theta_s)$, and $\ell(1-\cos\theta_s)/h = \theta_s - \sin\theta_s$. The curve is a **cycloid**.

Example 2.3
Find the minimum surface of revolution of the curve $y = y(x)$, rotating about the x axis, where $y_0 = y(x_0)$ and $y_1 = y(x_1)$. The surface is obtained from the minimum of the functional

$$I[y] = 2\pi\int_{x_0}^{x_1} y\sqrt{1+(y')^2}\, dx. \tag{2.36}$$

Equation (2.29) leads to

$$y = c\sqrt{1+y'^2}. \tag{2.37}$$

Therefore, we have

$$y = c\cos\left(\frac{x}{c} + c_1\right). \tag{2.38}$$

The integral constants c and c_1 are determined from the boundary conditions. The curve is called a **catenary**.

Potential Energies
Most functionals whose Euler differential equations are equations of equilibrium in mechanics equal the mechanical potential energies of bodies. The potential energy is the sum of the elastic

strain energy increase and the potential energy of an applied load which is the negative of the work done by the applied load.

Consider a string with length ℓ as shown in Fig. 2.5a. The string is in tension T. When a vertical load f is applied, the string deforms by y from the original horizontal line (Fig. 2.5b). The line element of the string is dx before loading and $\sqrt{1+(y')^2}\,dx \approx \left[1+\frac{1}{2}(y')^2\right]dx$ after loading for small deformations.

The elastic strain energy increase is $\frac{1}{2}T(y')^2$ per unit original length of the string. The potential energy of load is fy per unit original length of the string. The functional to be minimized is

$$I[y] = \frac{T}{2}\int_0^\ell (y')^2\,dx - \int_0^\ell f y\,dx \tag{2.39}$$

The Euler equation becomes

$$Ty'' + f = 0 \tag{2.40}$$

When (2.40) is solved with boundary conditions $y = 0$ at $x = 0$ and $x = \ell$, we have

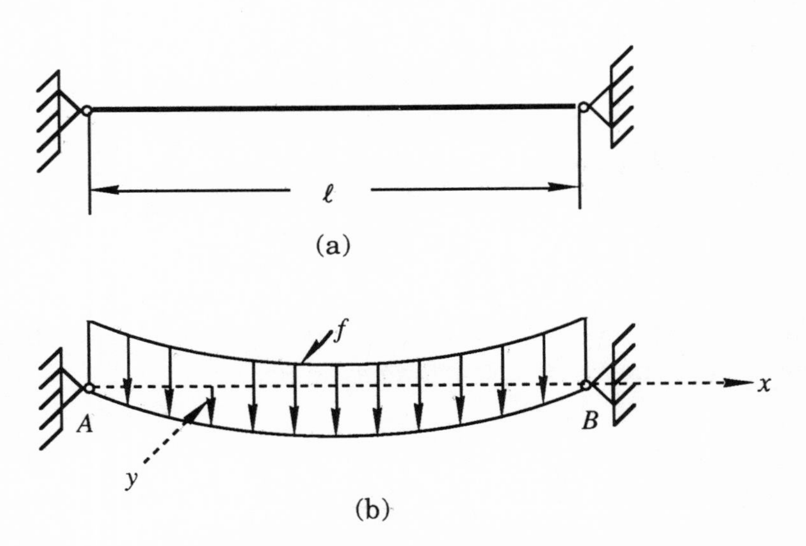

Figure 2.5(a) and (b). String under load f.

$$y = \frac{f}{2T}(\ell - x)x \tag{2.41}$$

Problems

2.1 A dislocation line $y = y(x)$ is lying between the atomic potential valleys at $y = \pm h$ (see Fig. P2.1). The energy of the dislocation per unit length is assumed to be $E(y) = E_0(1 + \cos(\pi y/2h))$. Find the dislocation configuration. (For an extended problem, see T. Mori and M. Kato, "Asymptotic form of activation energy for double-kink formation in a dislocation in a one-dimensional periodic field," *Philosophical Magazine A*, 1981, Vol, 43, No. 6, pp.1315-1320.)

Hint: $I[y] = \int_{-\infty}^{\infty} E(y)\sqrt{1 + y'^2}\,dx$ with the boundary condition $y = 0$ at $x = 0$, $y' = 0$ at $y = \pm h$.

Answer: $x = \dfrac{2h}{\pi}\displaystyle\int_0^z \frac{dz}{\sqrt{2\cos z + \cos^2 z}}$ for $x \geq 0$, $z = \dfrac{\pi y}{2h}$.

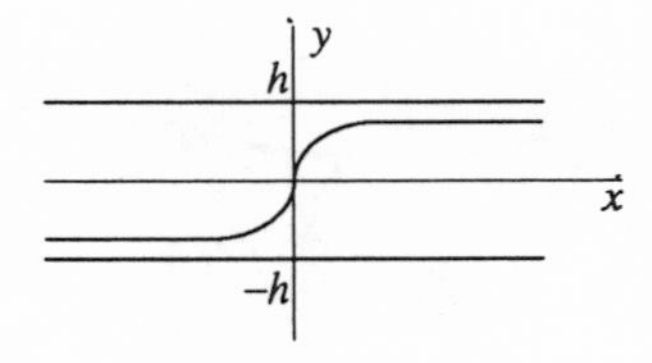

Figure P2.1. Dislocation line lying between the potential valleys.

2.2 Derive (2.35) and (2.38). Plot the cycloid with the coordinates x/a and y/a.

2.3 Sketch the stationary curve of the following functional

$$I[y] = \int_{-1}^{1} \sqrt{y[1 + (y')^2]}\,dx$$

with boundary conditions

$$y(-1) = 1 \qquad \text{and} \qquad y(1) = 1.$$

2.4 On a given surface, find the shortest wave (**geodesic curve**) between two points. The surface is given by the parametric representation $x = x(u, v)$, $y = y(u, v)$, and $z = z(u, v)$ in rectangular coordinates x, y, and z. The length I of a curve on the surface defined by the equation $v = v(u)$ between u_0 and u_1 is given by the integral

$$I[v] = \int_{u_0}^{u_1} \sqrt{e + 2fv' + g(v')^2}\, dx$$

where
$$e = x_u^2 + y_u^2 + z_u^2$$
$$g = x_v^2 + y_v^2 + z_v^2$$
$$f = x_u x_v + y_u y_v + z_u z_v$$

2.5 A line element ds in a non-Euclidean space is written as

$$ds = \sqrt{g_{ij} dx_i dx_j}$$

where g_{ij} are functions of $x_1, x_1, \ldots, x_n$. Derive the differential equations for geodesic curves (shortest distance between two points) of this space.

2.6 Consider a cone with angle α (see Fig. P2.6).

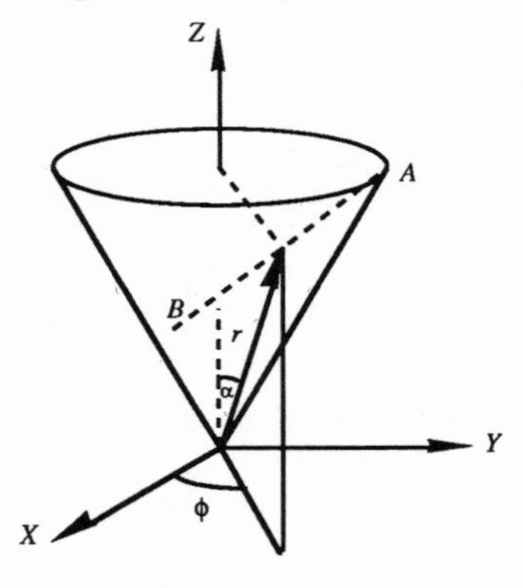

Figure P2.6. Cone with angle α.

The equation for the cone is $\theta = \alpha$ in the polar coordinate system. Find the equation of the geodesic between two points A and B whose polar coordinates are (r_1, α, ψ_1) and (r_2, α, ψ_2), respectively.

3

Ritz's Method

Let us consider the problem to minimize the functional

$$I[y] = \int_0^\ell (py'^2 + qy^2 + 2fy)dx \tag{3.1}$$

with the boundary condition

$$y(0) = y_0, \qquad y(\ell) = y_1. \tag{3.2}$$

We have **direct methods** to solve the problem rather than solving the associated Euler equation. **Ritz's method** is one of the direct methods.

We assume that the stationary function satisfying the boundary conditions (3.2) is

$$y_n(x) = \frac{x}{\ell}y_1 + \frac{\ell - x}{\ell}y_0 + \sum_{k=1}^{n} a_k \sin\frac{k\pi x}{\ell}. \tag{3.3}$$

Then

$$\begin{aligned}
I[y_n] = {} & \sum_{k,h=1}^{n} a_k a_h \int_0^\ell \left(p\frac{kh\pi^2}{\ell^2}\cos\frac{k\pi x}{\ell}\cos\frac{h\pi x}{\ell} + q\sin\frac{k\pi x}{\ell}\sin\frac{h\pi x}{\ell} \right)dx \\
& + 2\sum_{k=1}^{n} a_k \int_0^\ell \left\{ p\left(\frac{y_1 k\pi}{\ell^2} - \frac{y_0 k\pi}{\ell^2} \right)\cos\frac{k\pi x}{\ell} + q\left(\frac{xy_1}{\ell} + \frac{\ell - x}{\ell}y_0 \right)\sin\frac{k\pi x}{\ell} \right. \\
& \left. + f\sin\frac{k\pi x}{\ell} \right\} dx \\
& + \int_0^\ell \left\{ p\left(\frac{y_1^2}{\ell^2} + \frac{y_0^2}{\ell^2} - 2\frac{y_0 y_1}{\ell^2} \right) + q\left[\frac{x^2 y_1^2}{\ell^2} + \frac{(\ell - x)^2}{\ell^2}y_0^2 + 2\frac{x(\ell - x)}{\ell^2}y_0 y_1 \right] \right. \\
& \left. + 2f\left(\frac{xy_1}{\ell} + \frac{\ell - x}{\ell}y_0 \right) \right\} dx.
\end{aligned} \tag{3.4}$$

We can see that the value of I depends on the choice of the coefficients in (3.3). Therefore, I is an ordinary function of a_1, a_2, a_3, ..., after completion of the integrals in (3.4). Therefore, the stationary conditions are

$$\frac{\partial I}{\partial a_1}=0,\ \frac{\partial I}{\partial a_2}=0,\ldots,\ \frac{\partial I}{\partial a_n}=0 \tag{3.5}$$

which lead to

$$\sum_{h=1}^{n} a_h \int_0^{\ell}\left(p\frac{kh\pi^2}{\ell^2}\cos\frac{k\pi x}{\ell}\cos\frac{h\pi x}{\ell}+q\sin\frac{k\pi x}{\ell}\sin\frac{h\pi x}{\ell}\right)dx$$
$$+\int_0^{\ell}\left\{p\left(\frac{y_1 h\pi}{\ell^2}-\frac{y_0 h\pi}{\ell^2}\right)\cos\frac{h\pi x}{\ell}+q\left(\frac{xy_1}{\ell}+\frac{\ell-x}{\ell}y_0\right)\sin\frac{h\pi x}{\ell}+f\sin\frac{h\pi x}{\ell}\right\}dx=0,$$
$$k=1,2,3,\ldots,n. \tag{3.6}$$

The system of equations in (3.6) consists of equations for determining a_k. If n is large enough, we assume that y_n is sufficiently close to the stationary function y. This is Ritz's method. In this example, $\sin(k\pi x/\ell)$ is taken as the **fundamental function.** It is possible to take $(\ell-x)x^k$ instead of $\sin(k\pi x/\ell)$. All continuous functions having piecewise continuous derivatives satisfying (3.2) are admissible comparison functions in this problem.

If any continuously differentiable function y, which satisfies the **homogeneous boundary conditions** $y(0)=0$ and $y(\ell)=0$, is uniformly approximated by a linear combination of finite number of $w_1, w_2,\ldots, w_n$,

$$y_n=\sum_{k=1}^{n} a_k w_k, \tag{3.7}$$

the system of $w_1, w_2,\ldots, w_n$, is called the **complete system of functions**, and w_k is called a **fundamental function.** Examples of the complete system of functions are

$$w_k=(\ell-x)x^k,\quad w_k=\sin\frac{k\pi x}{\ell}. \tag{3.8}$$

The comparison function for (3.1) with (3.2) is taken as (3.3) that is

$$y=\frac{x}{\ell}y_1+\frac{\ell-x}{\ell}y_0+\sum_{k=1}^{\infty} a_k w_k.$$

Example 3.1
Minimize the functional

$$I[\psi]=\iint_R\left[\left(\frac{\partial\psi}{\partial x}\right)^2+\left(\frac{\partial\psi}{\partial y}\right)^2\right]dx\,dy \tag{3.9}$$

with the **subsidiary condition**

$$H[\psi]=\iint_R \psi^2\,dx\,dy=1 \tag{3.10}$$

and $\psi=0$ at the boundary of R, where R is the rectangular domain $0 \le x \le a$ and $0 \le x \le b$.

Assume the solution as

$$\psi=\sum_{m,n=1}^{\infty} a_{mn}\sin\frac{m\pi x}{a}\sin\frac{n\pi y}{b}. \tag{3.11}$$

Then

$$I=\pi^2\frac{ab}{4}\sum_{m,n=1}^{\infty} a_{mn}^2\left(\frac{m^2}{a^2}+\frac{n^2}{b^2}\right),$$

$$H=\frac{ab}{4}\sum_{m,n=1}^{\infty} a_{mn}^2.$$

The coefficients a_{mn} are determined from

$$\frac{\partial}{\partial a_{mn}}[I+\lambda(H-1)]=0,\quad \frac{\partial}{\partial\lambda}[I+\lambda(H-1)]=0 \tag{3.12}$$

where λ is Lagrange's multiplier. The equations in (3.12) are evaluated as

$$\left[\pi^2\left(\frac{m^2}{a^2}+\frac{n^2}{b^2}\right)+\lambda\right]a_{mn}=0,\quad (m,\ n=1,2,\ldots) \tag{3.13}$$

and

$$\frac{ab}{4}\sum_{m,n=1}^{\infty} a_{mn}^2=1.$$

We take all $a_{mn} = 0$ except a_{11}. Then $a_{11} = \pm 2/\sqrt{ab}$. The solution is

$$\psi = \frac{2}{\sqrt{ab}} \sin\frac{\pi x}{a} \sin\frac{\pi y}{b}. \tag{3.14}$$

Example 3.2
A solid bar is subjected to a given torsional moment M at the end $z = \ell$ (see Fig. 3.1). Find the stress field and the twisting angle. The stress components are denoted by τ_{zx} and τ_{zy}. Hooke's law is

$$\tau_{zx} = G\left(\frac{\partial w}{\partial x} + \frac{\partial u}{\partial z}\right), \qquad \tau_{zy} = G\left(\frac{\partial w}{\partial y} + \frac{\partial v}{\partial z}\right) \tag{3.15}$$

The boundary conditions are

$$\begin{aligned} \tau_{zx} &= 0 \ \text{ at } \ x = \pm a, \\ \tau_{zy} &= 0 \ \text{ at } \ y = \pm b, \end{aligned} \tag{3.16}$$

and

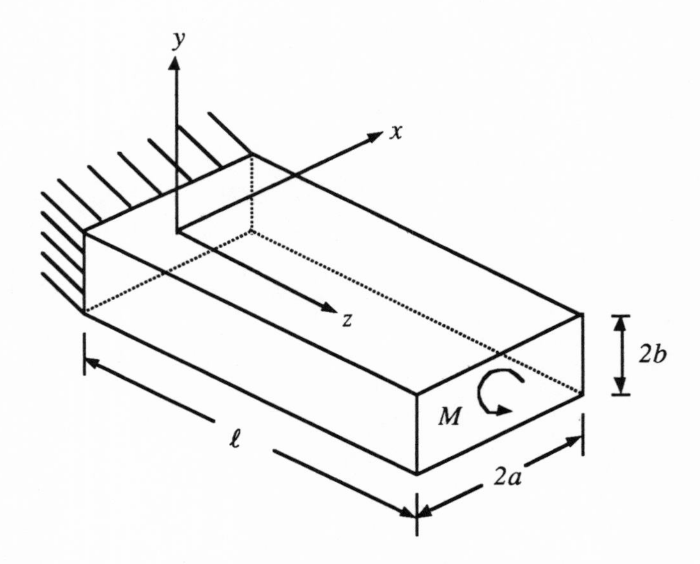

Figure 3.1. Torsional bar with moment M.

$$M = \int_{-a}^{a}\int_{-b}^{b}\left(\tau_{zy}x - \tau_{zx}y\right)dx\,dy, \text{ at } z=0 \tag{3.17}$$

where u, v, and w are the displacement components in the x, y, and z directions, respectively.

The total potential energy of the system is

$$I[u,v,w] = \frac{G}{2}\int_{-a}^{a}\int_{-b}^{b}\int_{0}^{\ell}\left[\left(\frac{\partial w}{\partial x}+\frac{\partial u}{\partial z}\right)^2+\left(\frac{\partial w}{\partial y}+\frac{\partial v}{\partial z}\right)^2\right]dx\,dy\,dz$$

$$-\frac{M}{4ab}\int_{-a}^{a}\int_{-b}^{b}\frac{1}{2}\left(\frac{\partial v}{\partial x}-\frac{\partial u}{\partial y}\right)_{z=\ell}dx\,dy. \tag{3.18}$$

The first integral is the elastic strain energy and the second integral is the potential energy of moment M.

Let us consider a minimum problem of (3.18) with the boundary conditions

$$u = v = w = 0 \text{ at } z = 0. \tag{3.19}$$

Since (3.18) has a positive definite quadratic form with respect to the highest derivatives of the arguments, $I[u,v,w]$ has a minimum value to minimize I. For a given M, the stationary functions u, v, and w are exact displacements realized by the given moment.

As a first approximation, we chose

$$u = -\theta zy, \quad v = \theta zx, \quad w = 0 \tag{3.20}$$

which are taken from the solution for a circular cylinder. θ is the angle of twist angle per unit length of the bar. The value of θ is chosen so that I takes a minimum value under the restriction of the choice of the form given by (3.20). When (3.20) is substituted into (3.18) we have

$$I = \frac{2G\theta^2\ell}{3}ab\left(a^2+b^2\right) - M\theta\ell. \tag{3.21}$$

Then $\partial I/\partial\theta = 0$ leads to

$$\theta = \frac{3M}{4Gab\left(a^2+b^2\right)}. \tag{3.22}$$

The corresponding minimum value of I is

$$I_{\min} = -\frac{3M^2\ell}{8\alpha\left(1+\alpha^2\right)Ga^4} \tag{3.23}$$

with $\alpha = b/a$.

A better approximation is obtained when we choose

$$u = -\theta zy, \quad v = \theta zx, \quad w = Axy. \tag{3.24}$$

When (3.24) is substituted into (3.18), we have

$$I = \frac{2G\ell}{3} ab\left[(A-\theta)^2 b^2 + (A+\theta)^2 a^2\right] - M\theta\ell. \tag{3.25}$$

The stationary conditions, $\partial I/\partial\theta = 0$ and $\partial I/\partial A = 0$, lead to

$$\theta = \frac{3M\left(1+\alpha^2\right)}{16\alpha^3 Ga^4}, \quad A = \frac{3\left(\alpha^2-1\right)M}{16\alpha^3 Ga^4} \tag{3.26}$$

with $\alpha = b/a$. The corresponding minimum value of I is

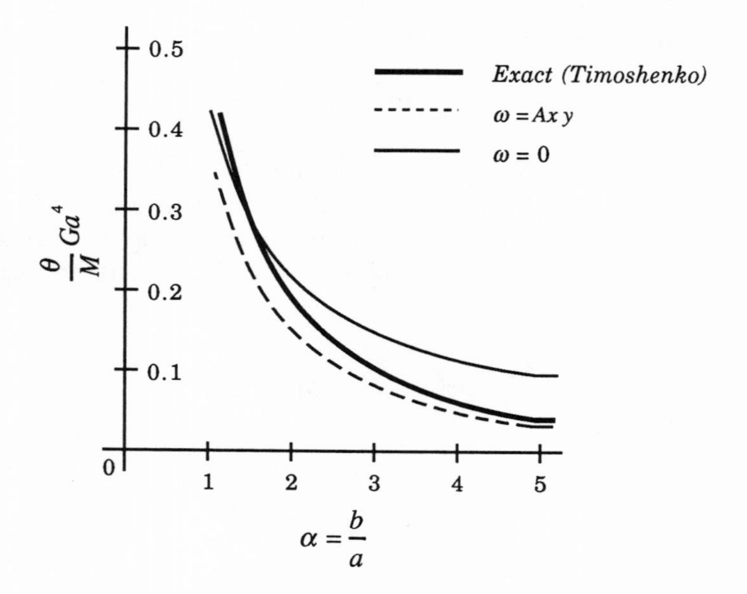

Figure 3.2. Torsional compliance against b/a in Fig. 3.1.

$$I_{\min} = -\frac{3(1+\alpha^2)M^2\ell}{32\alpha^3 G a^4}. \tag{3.27}$$

It is seen that (3.27) is smaller than (3.23). This indicates that the displacement components given by (3.24) are more realistic than those given by (3.20). The term M/θ is called the torsional rigidity. It is obtained from (3.22) for the first approximation and from (3.26) for the second approximation. The **torsional compliance** θ/M is plotted in Fig. 3.2 against α.

Ritz's Method by Mathematica

Symbolic manipulation codes (SMCs) on mainframes, workstations, and personal computers take symbolic approaches rather than traditional numerical approaches. Suppose we want to compute the integral

$$\int_0^1 x^2\, dx$$

on computer by any numerical method, the best one can obtain is

$$\int_0^1 x^2\, dx = 0.3333\ldots$$

up to the maximum precision of the computer. Because of the limited precision of the computer, even though the solution may be acceptable, it is not exact. On the other hand, SMCs yield

$$\int_0^1 x^2\, dx = \tfrac{1}{3}$$

which is the exact solution. SMCs do not approximate unless the user requests to do so. For some users, especially for those who have numerical programming experiences, the things SMCs can do are truly amazing. As long as numbers and equations are handled symbolically, they do not have to worry about round off errors and other numerical headaches.

SMCs can perform a variety of mathematical operations ranging from simple algebra to more complex calculus. Here are some sample operations:

$$\int x^2\,dx = \frac{1}{3}x^3 \text{ (constant neglected)},$$

$$\frac{d}{dx}\left(\frac{1}{3}x^3\right) = x^2,$$

$$\lim_{x\to 0}\frac{\sin x}{x} = 1,$$

$$(a+b)^2 = a^2 + 2ab + b^2,$$

and many more. We cannot perfom these operations by traditional numerical methods.

With the help of a SMC such as Mathematica, MACSYMA, REDUCE, and MAPLE, many tedious algebraic and calculus operations can be automated so that the user can concentrate on theoretical procedures rather than on numerical and symbolic manipulation details. Moreover, a SMC allows the user to perform more detailed analyses which, without it, one could never accomplish in finite time.

In the Appendix, Mathematica is used to perform Ritz's method to solve a few problems. It explains how to use Mathematica interactively step by step and how to program Mathematica to do all operations automatically. Once the user programs Mathematica, one can explore various problems without wasting time on algebraic details.

Kantorovich's Method

A hybrid of the direct and indirect method called **Kantorovich's method** takes a slightly different path from that of the Ritz's method introduced in the early part of this chapter. Consider a functional that has the form

$$I[y] = \int_t \int_x F(x,t,y)\,dx\,dt \tag{3.28}$$

with the boundary conditions

$$y|_{x=x_1} = y_1, \quad y|_{t=t_1} = y_{3,} \quad y|_{x=x_2} = y_2.$$

Suppose the trial function is of the form

$$y(x,t) \cong g_P(x) + \sum_{j=1} f_j(t) g_j^H(x) \tag{3.29}$$

where $g_P(x)$ satisfies the boundary conditions and $g_j^H(x)$ satisfies the homogeneous boundary conditions. At this moment, no condition is prescribed for $f_j(t)$. When this trial function is substituted into (3.28), a new reduced-degrees-of-freedom functional

$$I[f] = \int_t \overline{F}(f(t)) \tag{3.30}$$

is formed. Whence the Euler equation is derived and solved for $f_j(t)$.

Example 3.3
Use Kantrovich's method to extremize the functional

$$I[u] = \frac{1}{2}\int_0^1\int_0^\pi \left[\left(\frac{\partial u}{\partial x}\right)^2 + \left(\frac{\partial u}{\partial y}\right)^2\right] dx\, dy \tag{3.31}$$

with the boundary conditions

$$u\big|_{x=0} = u\big|_{x=\pi} = 0$$
$$u\big|_{y=0} = 0 \quad u\big|_{y=1} = \sin(x).$$

Assume the trial function

$$u(x,y) \cong x(x-\pi)f(y). \tag{3.32}$$

Notice that the trial function satisfies the boundary conditions in x. When this trial function is substituted into (3.31) and integrated over the x domain ($0 \le x \le \pi$), (3.31) is reduced to

$$I[f] = \frac{\pi^3}{6}\int_0^1 \left[f^2 + \frac{\pi^2}{10}(f')^2\right] dy. \tag{3.33}$$

The Euler equation of this functional is

$$f'' - \frac{10}{\pi^2} f = 0. \tag{3.34}$$

As for the boundary conditions for this differential equation, only the homogeneous boundary condition at $y = 0$ is used since the other condition is not homogenenous. The solution of this differential equation is then

$$f(y) = c\sinh\left(\frac{\sqrt{10}}{\pi}y\right). \tag{3.35}$$

The constant c is yet to be determined. When (3.35) is substituted back into (3.32), it becomes

$$u(x,y) \cong c\sinh\left(\frac{\sqrt{10}}{\pi}y\right)x(x-\pi). \tag{3.36}$$

To determine the constant c, the last boundary condition is used, that is,

$$u(x,1) = \sin(x) = c\sinh\left(\frac{\sqrt{10}}{\pi}\right)x(x-\pi). \tag{3.37}$$

Since there is no constant which satisfies the above equation exactly, the best fit constant is obtained by the following operations. Multiply $x(x - \pi)$ to both sides of (3.37) and integrate

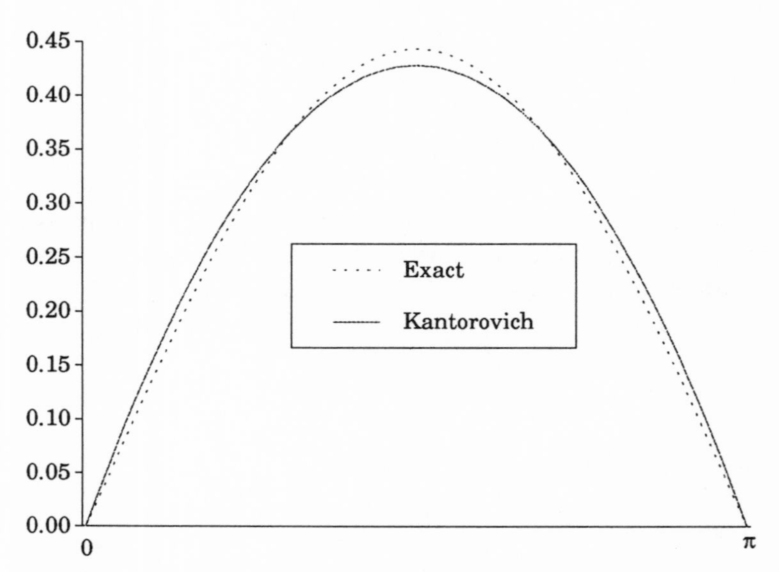

Figure 3.3. Kantorovich solution at $y = 0.5$.

them over the domain of x $(0 \le x \le \pi)$. The result is

$$\int_0^\pi \sin(x)x(x-\pi)dx = c\sinh\left(\frac{\sqrt{10}}{\pi}\right)\int_0^\pi \{x(x-\pi)\}^2\,dx$$
$$-4 = c\sinh\left(\frac{\sqrt{10}}{\pi}\right)\frac{\pi^5}{30} \tag{3.38}$$
$$c = \frac{-120}{\pi^5 \sinh\left(\sqrt{10}/\pi\right)}.$$

Therefore, an approximate solution for this problem is

$$u(x,y) \cong \frac{-120}{\pi^5 \sinh\left(\sqrt{10}/\pi\right)}\sinh\left(\frac{\sqrt{10}}{\pi}y\right)x(x-\pi). \tag{3.39}$$

The exact solution is

$$u(x,y) = \frac{1}{\sinh(1)}\sinh(y)\sin(x).$$

These two solutions are compared in Fig. 3.3.

In many cases, Kantorovich's method yields a better solution than the straight Ritz's method since part of the procedure is analytical.

The Least Squares Method

So far, Ritz's method has been used for the problems whose functional forms are given. In this section, Ritz's method is used only when the Euler equation is given. Consider the following problem. From Chapter 4, the Euler equation for the problem of uniform beam bending under a distributed load is

$$\frac{\partial^4 u}{\partial x^4} - \frac{1}{EI}f = 0 \tag{3.40}$$

where u is the deflection, E is Young's modulus, I is the moment of inertia of a transverse cross section, and f is the external load per unit length. Suppose the beam is simply supported, that is,

$$u|_{x=0} = u|_{x=l} = 0. \tag{3.41}$$

At this point, there are two choices: (1) find the functional whose Euler equation is (3.40), or (2) find a functional that serves the purpose of the functional found in (1).

***Path* (1):** The functional for the path (1) can be found by the following operations. First, multiply δu on both sides of (3.40) and integrate over the domain. The result is

$$\int_0^l \left[\frac{\partial^4 u}{\partial x^4}\delta u - \frac{1}{EI} f\delta u\right]dx = \frac{\partial^3 u}{\partial x^3}\delta u\bigg|_{x=l} - \frac{\partial^3 u}{\partial x^3}\delta u\bigg|_{x=0} - \frac{\partial^2 u}{\partial x^2}\delta u_x\bigg|_{x=l} + \frac{\partial^2 u}{\partial x^2}\delta u_x\bigg|_{x=0}$$
$$+\int_0^l \left[\frac{\partial^2 u}{\partial x^2}\delta u_{xx} - \frac{1}{EI} f\delta u\right]dx \quad (3.42)$$

Because of the boundary conditions $\delta u|_{x=0} = \delta u|_{x=l} = 0$. The natural boundary conditions are

$$\frac{\partial^2 u}{\partial x^2}\bigg|_{x=0} = \frac{\partial^2 u}{\partial x^2}\bigg|_{x=l} = 0. \quad (3.43)$$

The integral term in (3.42) is the variation of

$$I[u] = \int_0^l \left[\frac{1}{2}\left(\frac{\partial^2 u}{\partial x^2}\right)^2 - \frac{1}{EI} fu\right]dx. \quad (3.44)$$

To this functional, which is the same as (4.6), Ritz's method can be applied.

***Path* (2):** Suppose the trial function

$$u(x) \cong \bar{u}(x)$$

is assumed and substituted into (3.40). Because $\bar{u}(x)$ is an approximation to the exact solution $u(x)$, the right hand side of (3.40) is very likely nonzero. This nonzero function is the error due to approximation. (3.40) may be written as

$$\frac{\partial^4 \bar{u}}{\partial x^4} - \frac{1}{EI} f = \varepsilon(x). \quad (3.45)$$

When both sides are squared and integrated over the domain, a new functional

$$I[\bar{u}] = \int_0^l [\varepsilon(x)]^2 \, dx = \int_0^l \left(\frac{\partial^4 \bar{u}}{\partial x^4} - \frac{1}{EI} f \right)^2 dx \tag{3.46}$$

is formed. This functional, unlike the functional (3.44) that is the representation of the total potential energy, represents the total squared magnitude of the approximation error. The application of Ritz's method to this functional is known as the **least-squares method** to minimize the approximation error. In this problem, both paths yield the same solution.

Collocation Method

Perhaps this method is the most simple direct method which can also be considered as a variation of Ritz's method. This method also stems from the idea of minimizing the approximation error just as the least-squares method. However, whereas the least-squares method *minimizes* errors in the functional over the *entire* domain, the collocation method aims at the *vanishing* of errors in the original differential equation at selected points in the domain. Although these points (*collocation points*) can be selected arbitrarily, the accuracy of the solution depends heavily on the choice of these points.

Example 3.4

Consider the differential equation

$$\frac{\partial^2 u}{\partial x^2} + u = x^2, \quad u(0) = u(1) = 0. \tag{3.47}$$

In this example, the two-point collocation (excluding the boundary points) is used. Assume the trial function that satisfies the boundary conditions

$$u(x) \cong x(1-x)(a_1 + a_2 x). \tag{3.48}$$

The number of undetermined constants corresponds to the number of collocation points. When this trial function is substituted into

(3.47), the differential equation is rewritten as

$$(x^2 - x + 2)a_1 + (x^3 - x^2 + 6x - 2)a_2 = -x^2. \tag{3.49}$$

Suppose the collocation points are selected at $x = 1/3$ and $x = 2/3$. When these points are substituted into (3.49), a system of simultaneous equations is formed as below.

$$\begin{aligned} 48a_1 - 2a_2 &= -3 \\ 24a_1 + 25a_2 &= -6. \end{aligned} \tag{3.50}$$

Then the constants are

$$\begin{aligned} a_1 &= -\frac{29}{416}; \\ a_2 &= -\frac{72}{416}. \end{aligned} \tag{3.51}$$

Therefore, when these constants are substituted back into the trial function, the solution is

$$u(x) \cong -\frac{x(1-x)(29+72x)}{416}. \tag{3.52}$$

The error by this method is approximately -1.14196×10^{-4}. Although this method yields a very good approximation, extreme caution is advised since this method does not always yield the same solution

Case	x	Error
1	$\frac{1}{3}, \frac{2}{3}$	-1.14196×10^{-4}
2	$\frac{1}{6}, \frac{1}{3}$	-3.55506×10^{-3}
3	$\frac{2}{3}, \frac{5}{6}$	-2.86162×10^{-3}
4	$\frac{1}{8}, \frac{7}{8}$	-2.03460×10^{-3}

Table 3.1. Accuracy comparison with different collocation points.

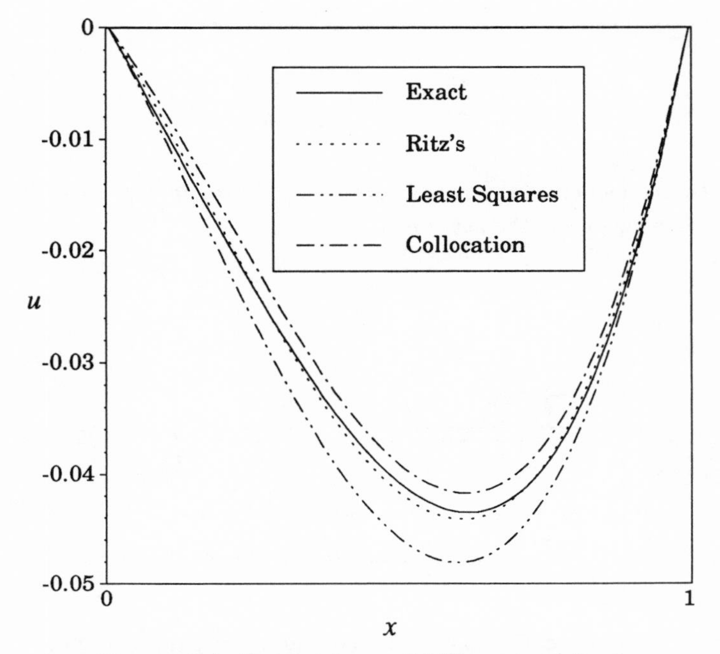

Figure 3.4. Comparison of different methods.

for the same trial function. The choice of collocation points determines the accuracy of solution as shown in Table 3.1. Fluctuations in accuracy become better as more points are selected. However, since there is no established procedure to select optimal points, one must rely on ones experience and intuition when not many points are used.

Finally, Fig.3.4 shows Ritz's, the least squares, and the collocation methods. They are compared based on (3.51) and the same trial function (3.52).

Problems

3.1 Minimize the functional defined in $0 \le x \le a$ and $0 \le y \le b$,

$$I[\psi] = \iint_R \left[\left(\frac{\partial \psi}{\partial x} \right)^2 + \left(\frac{\partial \psi}{\partial y} \right)^2 - 2\psi g \right] dx\, dy$$

with the boundary conditions

$$\psi = 0, \text{ on } x = 0 \text{ and } a, \quad y = 0 \text{ and } b,$$

where g is given by

$$g = \sum_{m,n=1}^{\infty} C_{mn} \sin\frac{m\pi x}{a} \sin\frac{n\pi y}{b}.$$

3.2 Find the deflection of the beam at $x = c$ as shown in Fig. P3.2, where the bending rigidity of the beam is EI.

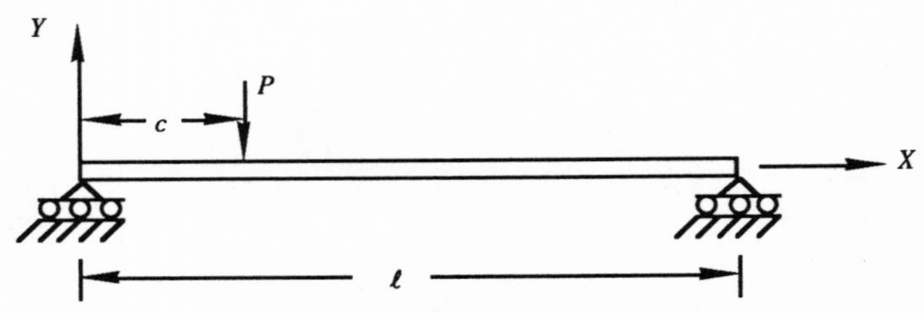

Figure P3.2. Beam under a concentrated force P point c.

3.3 Find the bending deflection of the circular ring shown in Fig. P3.3. The potential energy of the system is

$$I[w] = \frac{EI}{2r_0^3}\int_0^{2\pi}\left(\frac{d^2w}{d\theta^2} + w\right)^2 d\theta - P(w)_{\theta=\pi/2} - P(w)_{\theta=-\pi/2}.$$

Assume $w = \sum a_n \cos n\theta + \sum b_n \sin n\theta$.

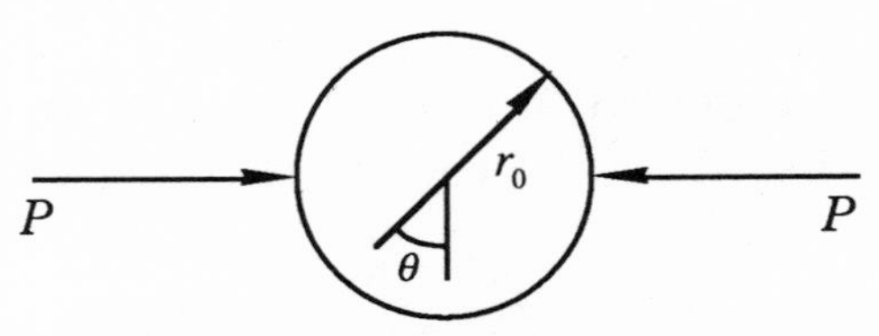

Figure P3.3. Circular beam with radius r_0 under a pair of force P.

3.4 Find the deflection of the plate, as shown in Fig. P3.4. The plate is simply supported along all the edges and is vertically subjected to a concentrated load P and compression N_x per unit length along the edges $x = 0$ and $x = a$. The potential energy is

$$I[w]=\frac{D}{2}\int_0^a\int_0^b\left[\left(\frac{\partial^2 w}{\partial x^2}+\frac{\partial^2 w}{\partial y^2}\right)^2-2(1-\nu)\left\{\frac{\partial^2 w}{\partial x^2}\frac{\partial^2 w}{\partial y^2}-\left(\frac{\partial^2 w}{\partial x\partial y}\right)^2\right\}\right]dx\,dy$$

$$-P(w)_{x=\zeta,\ y=\eta}-\frac{1}{2}\int_0^a\int_0^b N_x\left(\frac{\partial w}{\partial x}\right)^2 dx\,dy.$$

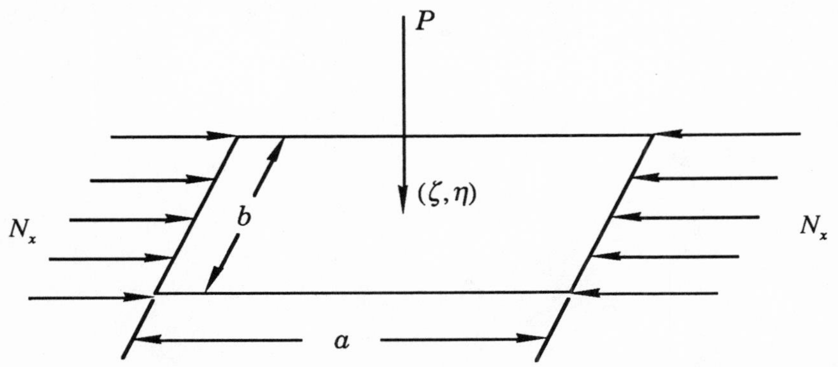

Figure P3.4. Bending of a plate with a concentrated force P and compressive plane force N_x.

3.5 Apply the direct method to finding the stationary function of

$$I[y]=\int_0^1\left[(y')^2+y^2+2xy\right]dx,$$

where $y=0$ at $x=0$ and 1.

4

The Euler Equation II

Higher Derivatives

The Euler differential equations with higher derivatives are obtained in an analogous manner for the variational problem of finding the stationary values of the functional

$$I[y]=\int_{x_0}^{x_1} F\left(x,y,y',\ldots y^{(n)}\right)dx, \tag{4.1}$$

where

$$\begin{aligned} y &= y_0 && \text{at } x=0 \\ y &= y_1 && \text{at } x=\ell \\ y' &= y_0' && \text{at } x=0 \\ y' &= y_1' && \text{at } x=\ell \\ &\vdots \\ y^{(n-1)} &= y_0^{(n-1)} && \text{at } x=0 \\ y^{(n-1)} &= y_1^{(n-1)} && \text{at } x=\ell, \end{aligned}$$

and F is a given function of the arguments $x,y,y',\ldots,y^{(n)}$. The first variation is

$$\delta I[y]=\int_{x_0}^{x_1}\left(F_y\delta y+F_{y'}\delta y'+\cdots+F_{y^{(n)}}\delta y^{(n)}\right)dx. \tag{4.2}$$

By carrying out integration by parts repeatedly, we can eliminate all the derivatives of y from the integral, transforming it into the form

$$\delta I[y]=\int_{x_0}^{x_1}\left[F_y-\left(F_{y'}\right)'+\left(F_{y''}\right)''-\cdots+(-1)^n\left(F_{y^{(n)}}\right)^{(n)}\right]\delta y\,dx. \tag{4.3}$$

The Euler equation becomes

$$[F]_y = F_y - \left(F_{y'}\right)' + \left(F_{y''}\right)'' - \cdots + (-1)^n \left(F_{y^{(n)}}\right)^{(n)} = 0. \tag{4.4}$$

The boundary terms resulting from the integration by parts vanish when

$$\delta y = 0, \quad \delta y' = 0, \ldots, \quad \delta y^{(n-1)} = 0 \tag{4.5}$$

at the boundaries or the coefficients of these variations at the boundaries are zero. This situation is illustrated by the following example.

Example 4.1

Minimize the mechanical potential energy of a *beam* with deflection y under applied force $f(x)$,

$$I[y] = \frac{EI}{2}\int_0^\ell (y'')^2\,dx - \int_0^\ell f\,y\,dx. \tag{4.6}$$

where EI is the bending rigidity of the beam span (Fig. 4.1).

The first variation becomes

$$\begin{aligned}
\delta I &= EI\int_0^\ell y''\delta y''dx - \int_0^\ell f\,\delta y dx \\
&= EI[y''\delta y']_0^\ell - EI\int_0^\ell y'''\delta y' dx - \int_0^\ell f\,\delta y dx \\
&= EI[y''\delta y']_0^\ell - EI[y'''\delta y]_0^\ell + EI\int_0^\ell y''''\delta y dx - \int_0^\ell f\,\delta y dx.
\end{aligned} \tag{4.7}$$

The Euler equation is obtained as

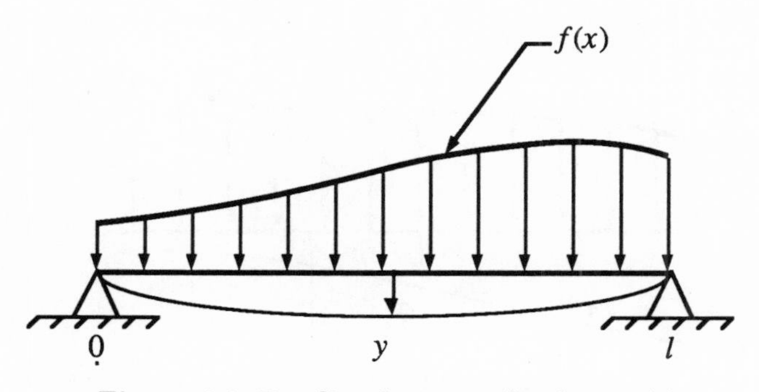

Figure 4.1. Bending beam under force $f(x)$.

$$EIy'''' - f = 0. \tag{4.8}$$

If the end of the beam at $x = 0$ is simply supported, the boundary conditions at $x = 0$ are

$$y = 0, \qquad y'' = 0. \tag{4.9}$$

The second condition is obtained because $\delta y'$ at $x = 0$ is arbitrary and therefore its coefficient y'' becomes zero. If the end at $x = 0$ is clamped, the boundary conditions at $x = 0$ are (Fig. 4.2)

$$y = 0, \qquad y' = 0. \tag{4.10}$$

For a free end (Fig. 4.3), δy and $\delta y'$ at $x = 0$ are arbitrary and their coefficients y'' and y''' must vanish. The boundary conditions at $x = \ell$ are obtained in the same manner.

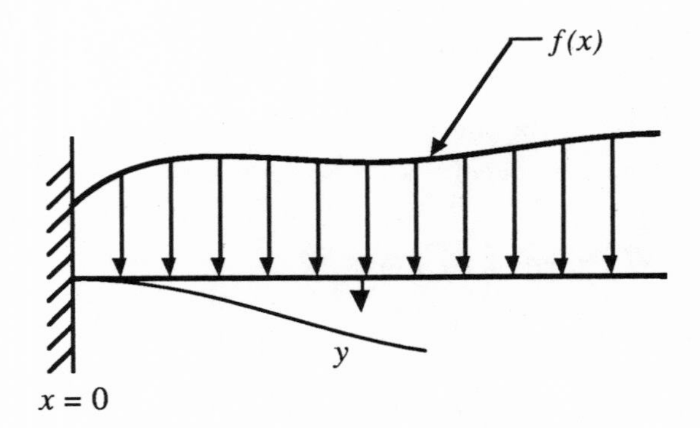

Figure 4.2. Clamped end.

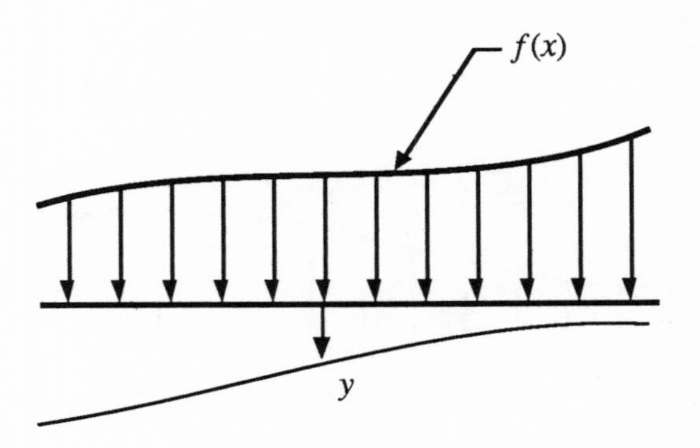

Figure 4.3. Free end.

The solution of (4.8) when f = constant and the two ends are simply supported is

$$EIy = \frac{f}{24}\left(x^4 - 2\ell x^3 + \ell^3 x\right). \tag{4.11}$$

Several Unknown Functions

Another extension is the problem of determining several functions $y(x)$, $z(x)$,... of x such that the functional

$$I[y,x,\ldots] = \int_{x_0}^{x_1} F(x,y,z,\ldots,y',z',\ldots)\,dx \tag{4.12}$$

is an extremum (or stationary value), where the values of the functions at the boundary points are given. The first variation must be zero,

$$\delta I = \int_{x_0}^{x_1} \left(F_y \delta y + F_{y'} \delta y' + \cdots + F_{z'} \delta z' + \cdots\right) dx = 0. \tag{4.13}$$

Integrating by parts lead to

$$\int_{x_0}^{x_1} \left\{ \left[F_y - \left(F_{y'}\right)' \right] \delta y + \left[F_z - \left(F_{z'}\right)' \right] \delta z + \cdots \right\} dx = 0, \tag{4.14}$$

where $\delta y = \delta z = 0$ at the boundary points. The Euler equations are

$$F_y - \left(F_{y'}\right)' = 0, \qquad F_z - \left(F_{z'}\right)' = 0, \ldots, \tag{4.15}$$

since $\delta y, \delta z, \ldots$ are independent arbitrary functions satisfying conditions for the fundamental lemma of the calculus of variation.

Example 4.2

There is a mechanical system characterized by n degrees of freedom with variable $q_1(t), q_2(t,), \ldots$ and $q_n(t)$, where t is time. The kinetic energy of the system is

$$T = P_{ij}\left(q_1, q_2, \ldots, q_n, t\right) \dot{q}_i(t) \dot{q}_j(t), \tag{4.16}$$

where the summation convention is used for repeated indices (e.g., $A_iB_i = A_1B_1 + A_2B_2 + \cdots + A_nB_n$) and where $\dot{q}_i = dq_i/dt$ and $P_{ij} = P_{ji}$. The potential energy of the system is

$$U = U(q_1, q_2, \ldots q_n, t). \tag{4.17}$$

P_{ij} and U are given functions of $q_1, q_2, \ldots, q_n, t$.

According to **Hamilton's principle**, the system is determined by the stationary condition for the functional

$$I[q_1, q_2, \ldots, q_n] = \int_{t_0}^{t_1} (T - U)\,dt \tag{4.18}$$

when the initial and end values of $q_1, q_2, \ldots, q_n$ are prescribed. The Euler equation is

$$\frac{\partial}{\partial q_i}(T - U) - \frac{d}{dt}\left(\frac{\partial T}{\partial \dot{q}_i}\right) = 0, \quad i = 1, 2, \ldots, n, \tag{4.19}$$

which is called **Lagrange's equation**. The function $T - U$ is called the **Lagrangian**.

As a special case, consider a unit mass connected with three springs, as shown in Fig. 4.4. The mass is located at the origin of the coordinates at rest. When it is displaced to point q_1, q_2, q_3 the necessary force has components $k_{ij}q_j$, in the q_i-direction. The potential energy is

$$U = \tfrac{1}{2} k_{ij} q_i q_j, \tag{4.20}$$

where $k_{ij} = k_{ji}$.

The kinetic energy is

$$T = \tfrac{1}{2} \dot{q}_i \dot{q}_i. \tag{4.21}$$

Lagrange's equation becomes

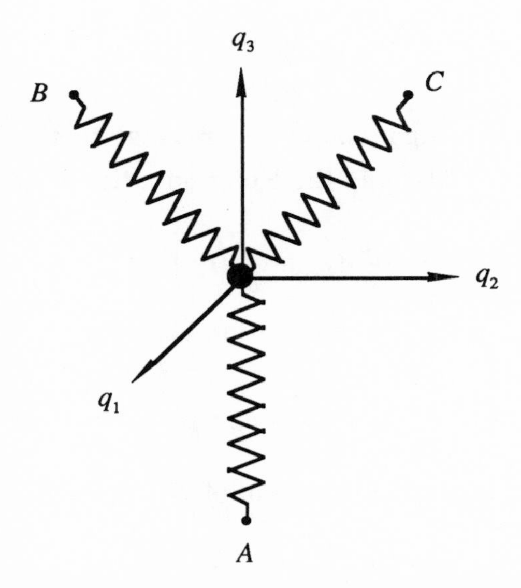

Figure 4.4. A mass is located at the origin of coordinates and connected by three springs.

$$k_{ij}q_j + \ddot{q}_i = 0, \quad i = 1,2,3, \tag{4.22}$$

which is equivalent to **Newton's equation**.

Several Independent Variables

Let us consider, for example, the problem of finding an extremum of the double integral

$$I[u] = \iint_G F(x, y, u, u_x, u_y)\,dx\,dy \tag{4.23}$$

over a given region of integration G by determining a suitable function u, which takes prescribed values on Γ, which is the boundary of G, where $\partial u/\partial x$ The first variation must be zero,

$$\delta I = \iint_G \left(F_u \delta u + F_{u_x} \delta u_x + F_{u_y} \delta u_y\right) dx\,dy = 0. \tag{4.24}$$

Integrating by parts leads to

$$\iint_G F_{u_x} \delta u_x \, dx \, dy = \int_\Gamma F_{u_x} n_x \delta u \, ds - \iint_G \frac{\partial}{\partial x}\left(F_{u_x}\right) \delta u \, dx \, dy, \tag{4.25}$$

where n_x is the x component of the normal to the boundary line element ds. Since $\delta u = 0$ on Γ, the Euler equation is obtained as

$$F_u - \frac{\partial}{\partial x}\left(F_{u_x}\right) - \frac{\partial}{\partial y}\left(F_{u_y}\right) = 0. \tag{4.26}$$

When $F = \frac{1}{2}\left(u_x^2 + u_y^2\right)$, the above equation becomes

$$\Delta u = u_{xx} + u_{yy} = 0.$$

Let us derive the second variation of (4.23). Substituting $u + \alpha\eta$ for u in (4.23), we have

$$I[\alpha] = \iint_G F\left(x, y, u + \alpha\eta, u_x + \alpha\eta_x, u_y + \alpha\eta_y\right) dx \, dy \tag{4.27}$$

We assume that (4.27) takes the extreme value when $\alpha = 0$. The first variation corresponds to $dI(\alpha)/d\alpha = 0$ and the second variation to $d^2 I(\alpha)/d\alpha^2$. Now,

$$\left\{\frac{d}{d\alpha} I(\alpha)\right\}_{\alpha=0} = \iint_G \left(F_u \eta + F_{u_x} \eta_x + F_{u_y} \eta_y\right) dx \, dy \tag{4.28}$$

and

$$\begin{aligned} \left\{\frac{d^2}{d\alpha^2} I(\alpha)\right\}_{\alpha=0} = \iint_G & \Big[\left(F_{uu}\eta^2 + F_{uu_x}\eta_x\eta + F_{uu_y}\eta_y\eta\right) \\ & + \left(F_{u_x}\eta\eta_x + F_{u_xu_x}\eta_x^2 + F_{u_xu_y}\eta_y\eta_x\right) \\ & + \left(F_{u_yu}\eta\eta_y + F_{u_yu_x}\eta_x\eta_y + F_{u_yu_y}\eta_y^2\right)\Big] dx \, dy. \end{aligned} \tag{4.29}$$

The Taylor expansion of $I(\alpha)$ in the neighborhood of $\alpha = 0$ is

$$I(\alpha) = I(0) + \left\{\frac{d}{d\alpha} I(\alpha)\right\}_{\alpha=0} \alpha + \frac{1}{2}\left\{\frac{d^2}{d\alpha^2} I(\alpha)\right\}_{\alpha=0} \alpha^2, \tag{4.30}$$

where only three terms are considered because α is infinitesimally small. The third term in (4.30) is the second variation.

The conditions necessary for (4.23) to have a minimum value are

$$\left\{\frac{d}{d\alpha}I(\alpha)\right\}_{\alpha=0}=0, \qquad \left\{\frac{d^2}{d\alpha^2}I(\alpha)\right\}_{\alpha=0}>0. \tag{4.31}$$

The conditions necessary for (4.23) to have a maximum value are

$$\left\{\frac{d}{d\alpha}I(\alpha)\right\}_{\alpha=0}=0, \qquad \left\{\frac{d^2}{d\alpha^2}I(\alpha)\right\}_{\alpha=0}<0. \tag{4.32}$$

The functional (4.23) has a saddle point at $\alpha=0$ when

$$\left\{\frac{d}{d\alpha}I(\alpha)\right\}_{\alpha=0}=0, \qquad \left\{\frac{d^2}{d\alpha^2}I(\alpha)\right\}_{\alpha=0}=0. \tag{4.33}$$

Example 4.3

For a *membrane* at rest, the potential energy is proportional to the change of area, and the proportionality factor is known as the tension T. Let $u(x,y)$ be the deflection. The change in area is

$$\begin{aligned}\sqrt{1+u_x^2+u_y^2}\,dx\,dy-dx\,dy &\approx \left[1+\tfrac{1}{2}\left(u_x^2+u_y^2\right)\right]dx\,dy-dx\,dy\\ &=\tfrac{1}{2}\left(u_x^2+u_y^2\right)dx\,dy.\end{aligned}$$

The internal energy due to tension T is

$$E_1=\iint_G \tfrac{1}{2}T\left(u_x^2+u_y^2\right)dx\,dy.$$

The external energy due to the applied load $f(x,y)$ is

$$E_2=-\iint_G f(x,y)u(x,y)\,dx\,dy.$$

Therefore, the total potential energy is

$$I[u]=\iint_G\left[\tfrac{1}{2}T\left(u_x^2+u_y^2\right)-fu\right]dx\,dy. \tag{4.34}$$

A mechanical system with the potential energy $I[u]$ is in equilibrium if and only if the potential energy is stationary. In order for the equilibrium to be stable, it is, moreover, necessary for the stationary value of $I[u]$ in equilibrium to be minimum. Thus the displacement $u(x,y)$ in the equilibrium position is that function which minimizes I. According to the variational principle, u is the solution of the Euler equation. The boundary condition is

$$u = 0 \quad \text{on } \Gamma, \tag{4.35}$$

where Γ is the boundary of G. The Euler equation becomes

$$T\left(\frac{\partial^2 u}{\partial x^2} + \frac{\partial^2 u}{\partial y^2}\right) + f = 0. \tag{4.36}$$

Particularly, if G is a circle with radius a and f = constant, the differential equation (4.36) becomes

$$\frac{T}{r}\frac{d}{dr}\left(r\frac{du}{dr}\right) + f = 0. \tag{4.37}$$

By integrating, we have

$$u = -\frac{fr^2}{4T} + c_2,$$

where $c_2 = fa^2/4T$ because $u = 0$ at $r = a$. Thus, the stationary function of (4.34) is

$$u = \frac{f}{4T}\left(a^2 - r^2\right). \tag{4.38}$$

The differential equation (4.36) is called **Poisson's differential equation**. When $f = 0$, it is called the **Laplace equation**. When u is prescribed on Γ, the boundary value problem is called **Dirichlet's problem**.

Example 4.4
The deflection of an elastic *plate* is determined by the minimum problem of the functional

$$I[u] = \frac{D}{2}\iint_G \left[\left(u_{xx} + u_{yy}\right)^2 - 2(1-\nu)\left(u_{xx}u_{yy} - u_{xy}^2\right) \right] dx\, dy$$
$$- \int_{\Gamma_1} p(s) u\, ds - \int_{\Gamma_1} m(s) \frac{\partial u}{\partial n}\, ds - \iint_G f u\, dx\, dy, \quad (4.39)$$

where D is the bending rigidity, ν is Poisson's ratio, f is a given load on the plate surface G, and p and m are a given force and moment on part Γ_1 of Γ. The deflection u and the slope $\partial u/\partial n$ are prescribed on Γ_2, which is $\Gamma - \Gamma_1$, and n is the normal distance from the boundary line element ds (see Fig. 4.5).

The first derivative becomes

$$\delta I = D\iint_G \left[\left(u_{xx} + u_{yy}\right)\left(\delta u_{xx} + \delta u_{yy}\right) - (1-\nu)\left(\delta u_{xx}u_{yy} + u_{xx}\delta u_{yy} - 2u_{xy}\delta u_{xy}\right) \right] dx\, dy$$
$$- \int_{\Gamma_1} p\delta u\, ds - \int_{\Gamma_1} m\frac{\partial \delta u}{\partial n} ds - \iint_G f\delta u\, dx\, dy$$
$$= D\iint_G \left\{ \left[\left(u_{xx} + u_{yy}\right) - (1-\nu)u_{yy} \right]\delta u_{xx} + \left[\left(u_{xx} + u_{yy}\right) - (1-\nu)u_{xx} \right]\delta u_{yy} \right.$$
$$\left. + (1-\nu)u_{xy}\delta u_{xy} + (1-\nu)u_{xy}\delta u_{yx} \right\} dx\, dy - \int_{\Gamma_1} p\delta u\, ds - \int_{\Gamma_1} m\frac{\partial \delta u}{\partial n} ds$$
$$- \iint_G f\delta u\, dx\, dy.$$

Further integration by parts lead to

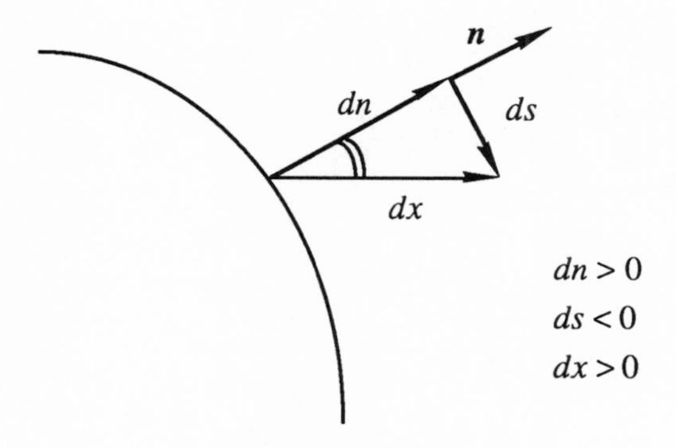

Figure 4.5. Geometry on the boundary.

$$\begin{aligned}
\delta I = D\int_{\Gamma}&\left[\left(u_{xx} + \nu u_{yy}\right)\delta u_x n_x + \left(u_{yy} + \nu u_{xx}\right)\delta u_y n_y\right.\\
&\left.+(1-\nu)u_{xy}\delta u_y n_x + (1-\nu)u_{xy}\delta u_x n_y\right] ds\\
&-D\iint_G\left[\left(u_{xxx} + \nu u_{yyx}\right)\delta u_x + \left(u_{yyy} + \nu u_{xxy}\right)\delta u_y\right.\\
&\left.+(1-\nu)u_{xyx}\delta u_y + (1-\nu)u_{xyy}\delta u_x\right] dx\, dy\\
&-\iint_G f\,\delta u\, dx\, dy - \int_{\Gamma_1} p\,\delta u\, ds - \int_{\Gamma_1} m\frac{\partial \delta u}{\partial n} ds
\end{aligned}\tag{4.40}$$

The second surface integral is transformed into

$$\begin{aligned}
-D\int_{\Gamma}&\left[\left(u_{xxx} + \nu u_{yyx}\right)n_x + \left(u_{yyy} + \nu u_{xxy}\right)n_y + (1-\nu)u_{xyx}n_y + (1-\nu)u_{xyy}n_x\right]\delta u\, ds\\
&+D\iint_G\left[\left(u_{xxxx} + \nu u_{yyxx}\right) + \left(u_{yyyy} + \nu u_{xxyy}\right) + 2(1-\nu)u_{xyxy}\right]\delta u\, dx\, dy.
\end{aligned}\tag{4.41}$$

where n_x and n_y are the components of the unit normal ds.

Along Γ, we have

$$\begin{aligned}
\delta u_x &= \frac{\partial}{\partial n}(\delta u)\frac{dn}{dx} + \frac{\partial}{\partial s}(\delta u)\frac{ds}{dx},\\
\delta u_y &= \frac{\partial}{\partial n}(\delta u)\frac{dn}{dy} + \frac{\partial}{\partial s}(\delta u)\frac{ds}{dy},
\end{aligned}\tag{4.42}$$

and

$$\frac{dn}{dx} = n_x, \quad \frac{ds}{dx} = -n_y, \tag{4.43}$$

as seen in Fig. 4.5, and

$$\frac{dn}{dy} = n_y, \quad \frac{ds}{dy} = n_x, \tag{4.44}$$

as seen in Fig. 4.6. Then

$$\begin{aligned}
\delta u_x &= \frac{\partial}{\partial n}(\delta u)n_x - \frac{\partial}{\partial s}(\delta u)n_y,\\
\delta u_y &= \frac{\partial}{\partial n}(\delta u)n_y + \frac{\partial}{\partial s}(\delta u)n_x.
\end{aligned}\tag{4.45}$$

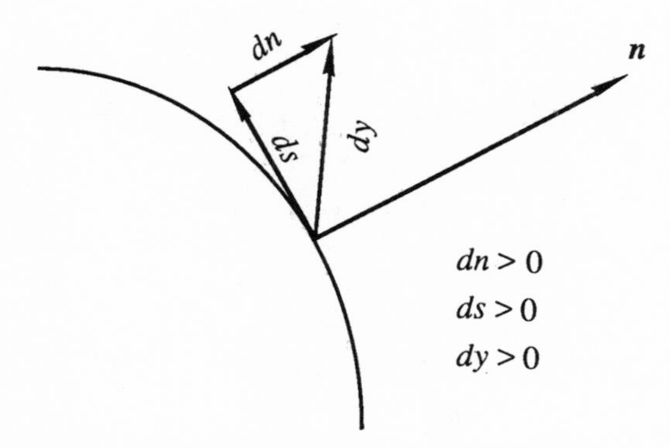

Figure 4.6. Geometry on the boundary.

Substituting (4.41) and (4.45) into (4.40) and applying integrations by parts to the terms containing $\partial(\delta u)/\partial s$, we have

$$\begin{aligned}\delta I &= \iint_G \left(D\Delta^2 u - f\right)\delta u\,dx\,dy - \int_{\Gamma_1} p\,\delta u\,ds - \int_{\Gamma_1} m\frac{\partial}{\partial n}(\delta u)\,ds\\ &+ D\int_\Gamma \Big[\left(u_{xx} + \nu u_{yy}\right)n_x n_x + 2(1-\nu)u_{xy}n_x n_y\\ &+ \left(u_{yy} + \nu u_{xx}\right)n_y n_y\Big]\frac{\partial}{\partial n}(\delta u)\,ds\\ &+ D\int_\Gamma \frac{\partial}{\partial s}\Big[\left(u_{xx} + \nu u_{yy}\right)n_x n_y + (1-\nu)u_{xy}\left(n_y^2 - n_x^2\right)\\ &- \left(u_{yy} + \nu u_{xx}\right)n_x n_y\Big]\delta u\,ds\\ &- D\int_\Gamma \left[\frac{\partial}{\partial x}\left(u_{xx} + u_{yy}\right)n_x + \frac{\partial}{\partial y}\left(u_{xx} + u_{yy}\right)n_y\right]\delta u\,ds. \end{aligned} \tag{4.46}$$

The Euler equations are

$$D\Delta\Delta u - f = 0 \qquad \text{in } G,$$

where

$$\Delta\Delta = \left(\frac{\partial^2}{\partial x^2} + \frac{\partial^2}{\partial y^2}\right)^2,$$

$$p(s) = -D\left[\frac{\partial}{\partial x}(u_{xx} + u_{yy})n_x + \frac{\partial}{\partial y}(u_{xx} + u_{yy})n_y\right]$$
$$+D\frac{\partial}{\partial s}\left[(u_{xx} + \nu u_{yy})n_x n_y + (1-\nu)u_{xy}(n_y^2 - n_x^2)\right.$$
$$\left.-(u_{yy} + \nu u_{xx})n_x n_y\right] \qquad \text{on } \Gamma_1,$$

and

$$m(s) = D\left\{(u_{xx} + \nu u_{yy})n_x n_x + 2(1-\nu)u_{xy}n_x n_y + (u_{yy} + \nu u_{xx})n_y n_y\right\} \quad \text{on } \Gamma.$$

Problems

4.1 Show that the Euler differential equation for

$$I[u] = \iint_G \left(u_{xx}u_{yy} - u_{xy}^2\right) dx\, dy$$

is identically zero.

4.2 Show that the Euler differential equation for

$$I[u] = \iint_G \frac{\left(u_{xx}u_{yy} - u_{xy}^2\right)}{\left(1 + u_x^2 + u_y^2\right)^{3/2}}\, dx\, dy$$

is identically zero. The integrand is the Gaussian curvature of the surface $z = u(x,y)$.

4.3 Do the functionals in Probs. 4.1 and 4.2 have minimum or maximum values when boundary values of derivatives of u are prescribed?

4.4 Derive the Euler equations with variables t for the functional

$$I[y] = \int_{x_0}^{x_1} F(x, y, y')\, dx = \int_{t_0}^{t_1} F^*(x, y, \dot{x}, \dot{y})\, dt = J[x, y]$$

where

$$F^*(x, y, \dot{x}, \dot{y}) = \dot{x}F(x, y, y'),$$
$$\dot{x} = \frac{dx}{dt}, \quad \dot{y} = \frac{dy}{dt}.$$

5

Boundary Conditions

Let us consider a functional

$$I=\int_{x_0}^{x_1}\left[p(y')^2+qy^2+2fy\right]dx=0. \tag{5.1}$$

where $p(x)$, $q(x)$, and $f(x)$ are continuous functions in the interval $x_0 \le x \le x_1$, $p(x)$ has continuous derivatives, and $y(x)$ has continuous second derivatives.

We are looking for a function y by which $I[y]$ takes a minimum value. Since no boundary conditions are given, the problem is called a **free boundary variational problem**.

The variation of (5.1) must be zero,

$$\delta I=\int_{x_0}^{x_1}2[py'\delta y'+qy\delta y+f\delta y]dx=0. \tag{5.2}$$

Integrating by parts leads to

$$\delta I=2[py'\delta y]_{x_0}^{x_1}+2\int_{x_0}^{x_1}\left[-(py')'+qy+f\right]\delta y dx=0. \tag{5.3}$$

Since $\delta y(x_1)$ and $\delta y(x_0)$ are arbitrary, the coefficient of δy at x_1 and x_0 must vanish,

$$py'=0 \text{ at } x_1 \text{ and } x_0. \tag{5.4}$$

Since δy is arbitrary in the domain $x_0 \le x \le x_1$, its coefficient must vanish in the domain,

$$-(py')'+qy+f=0 \quad \text{in } x_0 \le x \le x_1, \tag{5.5}$$

which is the Euler equation. The boundary conditions (5.4) are called the **natural boundary conditions**. The free boundary variational problem (5.1) is equivalent to the Euler equation (5.5) with the boundary conditions (5.4).

On the other hand, the variational problem (2.1) with the prescribed constraint boundary conditions (2.2) or with any other constraint conditions as discussed in the next section is called a **constraint variational problem**.

In mechanics, the boundary conditions given by (2.2) are displacement conditions and the conditions given by (5.4) are **natural conditions**. The Euler equation (5.5) is part of the natural conditions because (5.4) and (5.5) are derived in an equal base. In other words, (5.4) and (5.5) are called the natural conditions.

Let us consider a more general functional than (5.1),

$$I[y] = \int_{x_0}^{x_1} \left(py'^2 + qy^2 + 2fy\right)dx + h_1 y^2(x_1) - h_0 y^2(x_0), \tag{5.6}$$

where h_0 and h_1 are given constants. The corresponding natural conditions are

$$-(py')' + qy + f = 0 \quad \text{in } x_0 \le x \le x_1 \tag{5.7}$$

and

$$\begin{aligned} p(x_1)y'(x_1) + h_1 y(x_1) = 0, \\ p(x_0)y'(x_0) + h_0 y(x_0) = 0. \end{aligned} \tag{5.8}$$

It is seen that conditions (5.4) are obtained as a special case, $h_1 = h_0 = 0$, and conditions (2.2) as another special case, $h_1 = h_0 = \infty$, where $y_1 = y_0 = 0$.

Example 5.1

Let us consider again the beam problem considered in Example 4.1. Minimize the functional

$$I[y] = \frac{EI}{2}\int_0^{\ell}(y'')^2 dx - \int_0^{\ell} f y \, dx \tag{5.9}$$

with the boundary conditions

$$y = 0 \text{ at } x = 0 \text{ and } x = \ell. \tag{5.10}$$

The conditions in (5.10) are those for the simply supported beam. These conditions are the constraint boundary conditions.

When Ritz's method is applied, the trial function (comparison function) chosen must satisfy the constraint boundary conditions and does not need to satisfy the natural boundary conditions.

When the **indirect method** (solving the Euler equation) is employed, the first variation of (5.9) must be taken

$$\delta I = EI[y''\delta y']_0^\ell - EI[y'''\delta y]_0^\ell + EI\int_0^\ell y''''\delta y\,dx - \int_0^\ell f\,\delta y\,dx = 0 \qquad (5.11)$$

Since $\delta y = 0$ at $x = 0$ and $x = \ell$, the natural conditions are obtained as

$$EIy'''' - f = 0 \qquad \text{for } 0 < x < \ell. \qquad (5.12)$$

and

$$EIy'' = 0 \quad \text{at } x = 0 \text{ and } x = \ell \qquad (5.13)$$

In the indirect method, we have to solve the Euler equation (5.12) with all the boundary conditions (5.10) and (5.13).

Problems

5.1 Apply Ritz's method to the example problem 5.1. Choose the trial function

$$y = a_1x(x-\ell) + a_2x^2(x-\ell)^2$$

and compare this approximation with the exact solution (4.11) when f is constant.

5.2 Find the natural conditions for the stationary value of the functional

$$I[u] = \iint_G \left(u_x^2 + u_y^2 - 2fu\right)dx\,dy + \int_\Gamma \left[h(s)u^2 - 2g(s)u\right]ds,$$

where f, h, and g are given functions and Γ is the boundary of domain G. The resulting boundary-value problem is called **Neumann's problem** when $h = 0$.

5.3 State the variational method for a simply supported plate subjected to load f (the functional and the boundary conditions). Show the Euler equation and the natural boundary conditions.

5.4 An elastic body D is deformed by a given (prescribed) displacement on surface S_1 which is part of the surface S of D. The functional to minimize is

$$I[u_i] = \frac{1}{2}\int_D C_{ijkl} u_{k,l} u_{i,j} \, dD,$$

which is the elastic strain energy and u_i is the displacement. Find the Euler equations and the natural boundary conditions on $S - S_1$.

5.5 Find the Euler equation and the (natural) boundary conditions to minimize the functional

$$I[y] = \int_0^{\ell}\left(y'^2 + y^2 + 2xy\right)dx + \ell[y(0) - a] + \ell[y(\ell) - b]$$

where ℓ, a, and b are given constants.

5.6 Find the Euler equation and the initial conditions to minimize the Lagrangian

$$I[x, y, z] = \int_{t_0}^{t_1}\left[\frac{1}{2}m\left(\dot{x}^2 + \dot{y}^2 + \dot{z}^2\right) - \frac{1}{2}\kappa\left(x^2 + y^2 + z^2\right)\right]dt$$

where m is the mass of the particle and κ is the spring constant.

6

Subsidiary Conditions

Hitherto, variational problems subjected to constraint conditions were limited to boundary conditions in addition to continuity conditions except Example 3.1. In this section, we consider variational problems which are subjected to additional constraint conditions (**subsidiary conditions**).

Let us consider a problem to make the functional

$$I[y]=\int_{x_0}^{x_1}F(x,y,y')dx \tag{6.1}$$

stationary under the subsidiary condition

$$\int_{x_0}^{x_1}G(x,y,y')dx=\text{constant}=C \tag{6.2}$$

and the boundary conditions

$$y(x_0)=y_0, \quad y(x_1)=y_1, \tag{6.3}$$

where all necessary conditions for continuity are satisfied. This problem is called the **isoperimetric problem.**

Let us assume that $y(x)$ is a stationary function. A comparison function may be expressed as

$$\begin{aligned} y &= y(x)+\delta y \\ &= y(x)+\alpha_1\eta(x)+\alpha_2\zeta(x), \end{aligned} \tag{6.4}$$

where α_1 and α_2 are parameters, and $\eta(x)$ and $\zeta(x)$ are arbitrary functions of x having the necessary continuity condition and vanishing at the boundaries.

Substituting (6.4) into (6.1) and (6.2) gives

$$\begin{aligned} \phi(\alpha_1,\alpha_2)&=\int_{x_0}^{x_1}F(x,y+\alpha_1\eta+\alpha_2\zeta,y'+\alpha_1\eta'+\alpha_2\zeta')dx, \\ \psi(\alpha_1,\alpha_2)&=\int_{x_0}^{x_1}G(x,y+\alpha_1\eta+\alpha_2\zeta,y'+\alpha_1\eta'+\alpha_2\zeta')dx=C. \end{aligned} \tag{6.5}$$

Since ϕ takes a stationary value under $\psi = C$ when $\alpha_1 = 0$ and $\alpha_2 = 0$, the use of a Lagrange multiplier λ leads to making the functions $f + \lambda y$ stationary. The necessary conditions are

$$\left[\frac{\partial}{\partial \alpha_1}(\phi + \lambda\psi)\right]_{\alpha_1=\alpha_2=0} = 0,$$
$$\left[\frac{\partial}{\partial \alpha_2}(\phi + \lambda\psi)\right]_{\alpha_1=\alpha_2=0} = 0, \tag{6.6}$$

or

$$\int_{x_0}^{x_1}(F_y\eta + F_{y'}\eta')dx + \lambda\int_{x_0}^{x_1}(G_y\eta + G_{y'}\eta')dx = 0,$$
$$\int_{x_0}^{x_1}(F_y\zeta + F_{y'}\zeta')dx + \lambda\int_{x_0}^{x_1}(G_y\zeta + G_{y'}\zeta')dx = 0. \tag{6.7}$$

Further integration by parts leads to

$$\int_{x_0}^{x_1}\{[F]_y + \lambda[G]_y\}\eta\,dx = 0,$$
$$\int_{x_0}^{x_1}\{[F]_y + \lambda[G]_y\}\zeta\,dx = 0, \tag{6.8}$$

where

$$[F]_y = F_y - \frac{d}{dx}\left(F_{y'}\right),$$
$$[G]_y = G_y - \frac{d}{dx}\left(G_{y'}\right). \tag{6.9}$$

For arbitrarity η and ζ, the Euler equations are obtained as

$$[F]_y + \lambda[G]_y = 0. \tag{6.10}$$

Example 6.1

A uniform string with length C and density r is hanging between two points (x_0,y_0) and (x_1,y_1) (see Fig. 6.1). With the gravitational force acting in the y direction, find the shape of the string in equilibrium.

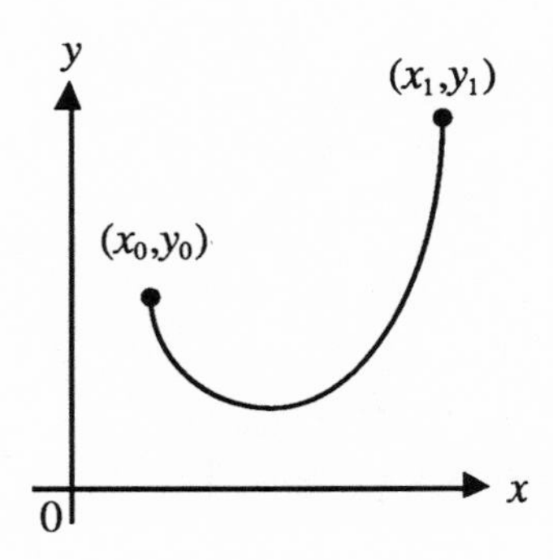

Figure 6.1. Hanging string.

The equilibrium state is determined from the minimum potential energy condition. The potential energy is

$$I[y] = \rho \int_{x_0}^{x_1} y\sqrt{1+y'^2}\, dx. \tag{6.11}$$

The length of the string is constant

$$\int_{x_0}^{x_1} \sqrt{1+y'^2}\, dx = C. \tag{6.12}$$

The Euler equation (6.10) becomes

$$\rho\sqrt{1+y'^2} - \rho\frac{d}{dx}\left(\frac{yy'}{\sqrt{1+y'^2}}\right) - \lambda\frac{d}{dx}\left(\frac{y'}{\sqrt{1+y'^2}}\right) = 0. \tag{6.13}$$

The boundary conditions are

$$y(x_0) = y_0, \quad y(x_1) = y_1. \tag{6.14}$$

Since $F+\lambda G$ does not explicitly contain x, (2.29) is applicable when F in (2.29) is replaced by $F + \lambda G$, where $F = \rho y\sqrt{1+(y')^2}$ and $G = \sqrt{1+(y')^2}$. Thus (2.29) leads to

$$\rho y\sqrt{1+y'^2} + \lambda\sqrt{1+y'^2} - \rho y\frac{y'^2}{\sqrt{1+y'^2}} - \lambda\frac{y'^2}{\sqrt{1+y'^2}} = \text{constant} = C_1 \tag{6.15.1}$$

or

$$\frac{\rho y + \lambda}{\sqrt{1+y'^2}} = C_1. \tag{6.15.2}$$

Then

$$\int \frac{dy}{\sqrt{\{(\rho y + \lambda)/C_1\}^2 - 1}} = \int dx \tag{6.16.1}$$

and

$$\frac{\rho x}{C_1} + C_2 = \cosh^{-1} \frac{\rho y + \lambda}{C_1}. \tag{6.16.2}$$

The three unknown constants C_1, C_2, and λ are determined from (6.14) and (6.12),

$$\rho y_0 + \lambda = C_1 \cosh\left(\frac{\rho x_0}{C_1} + C_2\right),$$

$$\rho y_1 + \lambda = C_1 \cosh\left(\frac{\rho x_1}{C_1} + C_2\right), \tag{6.17}$$

$$\frac{C_1}{\rho}\left[\sinh\left(\frac{\rho x_1}{C_1} + C_2\right) - \sinh\left(\frac{\rho x_0}{C_1} + C_2\right)\right] = C.$$

The function y defined by (6.16.1) is called the **catenary.** Let us consider a functional

$$I[y,z] = \int_{x_0}^{x_1} F(x,y,z,y',z')\,dx. \tag{6.18}$$

Let us find stationary function $y = y(x)$ and $z = z(x)$ with the boundary conditions

$$y(x_0) = y_0, \quad y(x_1) = y_1,$$
$$z(x_0) = z_0, \quad z(x_1) = z_1 \tag{6.19}$$

and a subsidiary condition

$$G(x, y, z, y', z') = 0. \tag{6.20}$$

This problem is different from the problem (6.1) to (6.3) in the sense that the subsidiary condition is given in the form of a differential equation. In this case, the Lagrange multiplier λ is taken as a function of x. Then, the new functional to be considered is

$$J[y, z, \lambda] = \int_{x_0}^{x_1} F(x, y, z, y', z')\,dx + \int_{x_0}^{x_1} \lambda G(x, y, z, y', z')\,dx. \tag{6.21}$$

Then

$$\begin{aligned} \delta J &= \int_{x_0}^{x_1} \left(F_y \delta y + F_{y'} \delta y' + F_z \delta z + F_{z'} \delta z'\right) dx \\ &\qquad + \int_{x_0}^{x_1} \lambda\left(G_y \delta y + G_{y'} \delta y' + G_z \delta z + G_{z'} \delta z'\right) dx \\ &= \int_{x_0}^{x_1} \left\{F_y - \left(F_{y'}\right)' + \lambda G_y - \left(\lambda G_{y'}\right)'\right\} \delta y\, dx \\ &\qquad + \int_{x_0}^{x_1} \left\{F_z - \left(F_{z'}\right)' + \lambda G_z - \left(\lambda G_{z'}\right)'\right\} \delta z\, dx \end{aligned} \tag{6.22}$$

since $\delta y = 0$ and $\delta z = 0$ at $x = x_0$ and at $x = x_1$.

The Euler equations are

$$\begin{aligned} F_y - \left(F_{y'}\right)' + \lambda G_y - \left(\lambda G_{y'}\right)' &= 0, \\ F_z - \left(F_{z'}\right)' + \lambda G_z - \left(\lambda G_{z'}\right)' &= 0, \end{aligned} \tag{6.23}$$

where $\lambda = \lambda(x)$.

When G does not contain y' and z', $Gy' = 0$ and $Gz' = 0$ in (6.23), but λ is still interpreted as a function of x.

The reason why λ is a function of x is obvious. The condition (6.20) is defined at any point of x. If the interval (x_0, x_1) is divided into n intervals with a sufficiently large n, condition (6.20) must hold in the n intervals, and, therefore, we need n Lagrange multipliers. If the intervals are infinitesimally small, the Lagrange multipliers are replaced by a single multiplier that is, however, a function of x.

Example 6.2
Minimize

$$I[\psi]=\tfrac{1}{2}\iint_G\left(\psi_x^2+\psi_y^2\right)dx\,dy \tag{6.24}$$

with the boundary condition

$$\psi=\psi_0(s) \qquad \text{on } \Gamma, \tag{6.25}$$

where

$$\psi_x=\frac{\partial\psi}{\partial x} \text{ and } \psi_y=\frac{\partial\psi}{\partial y}. \tag{6.26}$$

The last condition (6.26) can be interpreted as a condition similar to (6.20), assuming that ψ_x and ψ_y are independent quantities. Then our variational problem involves finding a stationary value of

$$I[\psi,\psi_x',\psi_y',\lambda_x,\lambda_y]=\iint_G\left[\tfrac{1}{2}\left(\psi_x^2+\psi_y^2\right)+\lambda_x\left(\psi_x-\frac{\partial\psi}{\partial x}\right)\right.$$
$$\left.+\lambda_y\left(\psi_y-\frac{\partial\psi}{\partial y}\right)\right]dx\,dy, \tag{6.27}$$

where λ_x and λ_y are the Lagrange multipliers. The first variation is zero,

$$\delta I=\iint_G\left[\psi_x\delta\psi_x+\psi_y\delta\psi_y+\lambda_x\delta\psi_x+\lambda_y\delta\psi_y-\lambda_x\frac{\delta\psi_x}{\partial x}-\lambda_y\frac{\delta\psi_y}{\partial y}\right.$$
$$\left.+\left(\psi_x-\frac{\partial\psi}{\partial x}\right)\delta\lambda_x+\left(\psi_y-\frac{\partial\psi}{\partial y}\right)\delta\lambda_y\right]dx\,dy=0. \tag{6.28}$$

Integration by parts leads to

$$\delta I = \iint_G \left[(\psi_x + \lambda_x)\delta\psi_x + (\psi_y + \lambda_y)\delta\psi_y + \left(\frac{\partial}{\partial x}\lambda_x + \frac{\partial}{\partial y}\lambda_y \right)\delta\psi \right.$$
$$\left. + \left(\psi_x - \frac{\partial\psi}{\partial x} \right)\delta\lambda_x + \left(\psi_y - \frac{\partial\psi}{\partial y} \right)\delta\lambda_y \right] dx\,dy = 0. \tag{6.29}$$

Then the natural conditions are

$$\psi_x + \lambda_x = 0, \qquad \psi_y + \lambda_y = 0,$$
$$\frac{\partial}{\partial x}\lambda_x + \frac{\partial}{\partial y}\lambda_y = 0, \qquad \psi_x - \frac{\partial\psi}{\partial x} = 0, \qquad \psi_y - \frac{\partial\psi}{\partial y} = 0. \tag{6.30}$$

It is easily seen that the conditions in (6.30) are equivalent to

$$\frac{\partial^2\psi}{\partial x^2} + \frac{\partial^2\psi}{\partial y^2} = 0. \tag{6.31}$$

The boundary condition $\psi = \psi(s)$ is also considered as a subsidiary condition, and it is included in I by a Lagrange multiplier λ defined on Γ. Then the new functional becomes

$$I[\psi_x, \psi_y, \psi, \lambda_x, \lambda_y, \lambda] = \iint_G \left[\tfrac{1}{2}(\psi_x^2 + \psi_y^2) + \lambda_x \left(\psi_x - \frac{\partial\psi}{\partial x} \right) \right.$$
$$\left. + \lambda_y \left(\psi_y - \frac{\partial\psi}{\partial y} \right) \right] dx\,dy + \int_\Gamma \lambda(\psi - \psi_0)\,ds. \tag{6.32}$$

Then the natural conditions are obtained as

$$\left. \begin{aligned} &\psi_x + \lambda_x = 0, \qquad \psi_y + \lambda_y = 0, \qquad \frac{\partial\lambda_x}{\partial x} + \frac{\partial\lambda_y}{\partial y} = 0 \\ &\psi_x - \frac{\partial\psi}{\partial x} = 0, \qquad \psi_y - \frac{\partial\psi}{\partial y} = 0, \end{aligned} \right\} \text{ in } G \tag{6.33}$$

and

$$\lambda - \lambda_x n_x - \lambda_y n_y = 0, \quad \psi - \psi_0 = 0 \quad \text{on } \Gamma. \tag{6.34}$$

The solution of the problem is obtained by solving the system of equations (6.33) and (6.34).

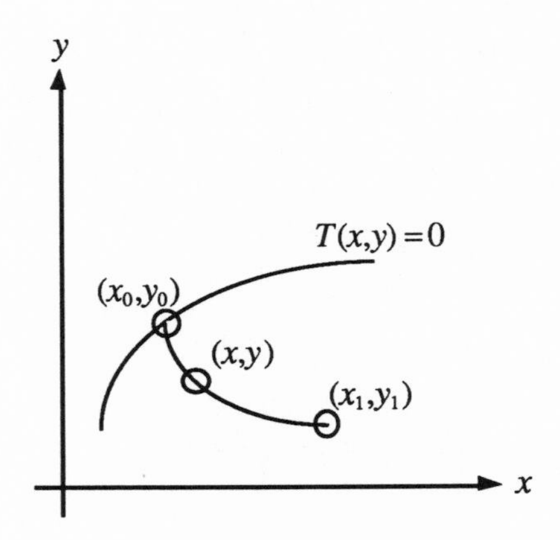

Figure 6.2. Curve $T(x,y)=0$ and point (x_0,y_0) are given. The stationary functions starts from the point and ends on the curve.

Example 6.3
Find a stationary function of

$$I[y]=\int_{x_0}^{x_1} F(x,y,y')\,dx, \tag{6.35}$$

where $y = y(x)$ is a curve connecting a point (x_0, y_0) and a point located on a curve (see Fig. 6.2),

$$T(x,y)=0. \tag{6.36}$$

By the use of a parameter t, point (x, y) is described by

$$x=x(t), \quad y=y(t), \tag{6.37}$$

where

$$t_0 \le t \le t_1, \tag{6.38}$$

$x(t_0) = x_0$, $x(t_1) = x_1$, $y(t_0) = y_0$ and $y(t_1) = y_1$. Then (6.35) is written as

$$I[y]=\int_{t_0}^{t_1} \dot{x}F\left(x,y,\frac{\dot{y}}{\dot{x}}\right)dt. \tag{6.39}$$

Since $y' = dy/dx = \dot{y}/\dot{x}$, $dx = \dot{x}\,dt$. The subsidiary condition (6.36) is

$$T\big(x(t_1), y(t_1)\big) = 0. \tag{6.40}$$

The new functional constructed from (6.39) and (6.40) becomes

$$I[x, y, \lambda] = \int_{t_0}^{t_1} \dot{x} F\left(x, y, \frac{\dot{y}}{\dot{x}}\right) dt + \lambda T\big(x(t_1), y(t_1)\big). \tag{6.41}$$

The first variation with respect to x and y becomes zero,

$$\delta I = \int_{t_0}^{t_1} \left[\delta\dot{x} F + \dot{x}\left(F_x \delta x + F_y \delta y + F_{\dot{x}} \delta\dot{x} + F_{\dot{y}} \delta\dot{y}\right)\right] dt$$
$$+ \lambda\left(T_x \delta x + T_y \delta y\right)_{t=t_1} + \delta\lambda T\big(x(t_1), y(t_1)\big) = 0. \tag{6.42}$$

The natural conditions are obtained by integrations by parts with respect to t, applied to $\delta\dot{x}$ and $\delta\dot{y}$,

$$\begin{aligned} \dot{x} F_x - \frac{d}{dt}\left(F + \dot{x} F_{\dot{x}}\right) &= 0, \\ \dot{x} F_y - \frac{d}{dt}\left(F + \dot{x} F_{\dot{y}}\right) &= 0, \end{aligned} \tag{6.43}$$

and for $t = t_1$

$$\begin{aligned} F + \dot{x} F_{\dot{x}} + \lambda T_x &= 0, \\ \dot{x} F_{\dot{y}} + \lambda T_y &= 0, \\ T &= 0, \end{aligned} \tag{6.44}$$

where $\delta x = 0$, $\delta y = 0$, at $t = t_0$.

When λ is eliminated from (6.44), we have

$$\frac{F + \dot{x} F_{\dot{x}}}{T_x} = \frac{\dot{x} F_{\dot{y}}}{T_y}. \tag{6.45}$$

The condition in (6.45) is called the **transversality condition.**

When point (x_0, y_0) is also located on another curve $T_0 = 0$, a relation similar to (6.45) holds.

As an example, consider the problem of finding the shortest distance between two curves defined by $x = y$ and $y^2 = x - 1$. The integrand F in expression (6.35) is

$$F = \sqrt{1 + \left(\frac{\dot{y}}{\dot{x}}\right)^2}, \tag{6.46}$$

and therefore,

$$\begin{aligned} F_{\dot{x}} &= \frac{-\dot{y}^2/\dot{x}^3}{\sqrt{1+(\dot{y}/\dot{x})^2}} \\ F_{\dot{y}} &= \frac{-\dot{y}/\dot{x}^2}{\sqrt{1+(\dot{y}/\dot{x})^2}}. \end{aligned} \tag{6.47}$$

The equations in (6.43) are obtained as

$$\begin{aligned} \frac{1}{\sqrt{1+(\dot{y}/\dot{x})^2}} &= \text{constant}, \\ \frac{\dot{y}/\dot{x}}{\sqrt{1+(\dot{y}/\dot{x})^2}} &= \text{constant}, \end{aligned} \tag{6.48}$$

that lead to

$$\frac{\dot{y}}{\dot{x}} = \text{const.} \tag{6.49}$$

or

$$y = c_1 + c_2 x.$$

Since $T = x - y$ at $t = t_1$, the transversality condition (6.45) becomes

$$1 = -\frac{\dot{y}}{\dot{x}}. \tag{6.50}$$

The transversality condition for $t = t_0$ is also expressed by (6.45), but $T = y^2 - x + 1$. Then the condition becomes

$$-1 = \frac{\dot{y}/\dot{x}}{2y}. \tag{6.51}$$

From (6.50) and (6.49) we have $c_2 = -1$. From (6.51), (6.49), and $y^2 - x + 1 = 0$, we have $c_1 = \frac{7}{4}$. Thus, we have the curve

$$y = \tfrac{7}{4} - x. \tag{6.52}$$

Problems

6.1 Find a two-dimensional figure taking a minimum circumference under a constant area.

6.2 Find a two-dimensional figure taking a maximum area under a constant circumference.

Hint: $$J[r,\theta] = \tfrac{1}{2}\int_{t_0}^{t_1} r^2\dot{\theta}\,dt + \lambda\int_{t_0}^{t_1}\sqrt{\dot{r}^2 + r^2\dot{\theta}^2}\,dt.$$

Answer: $$a^2 = b^2 + r^2 - 2br\cos(\theta - \alpha).$$

6.3 A column with length ℓ is subjected to a load P, as shown in Fig. P6.3. Find

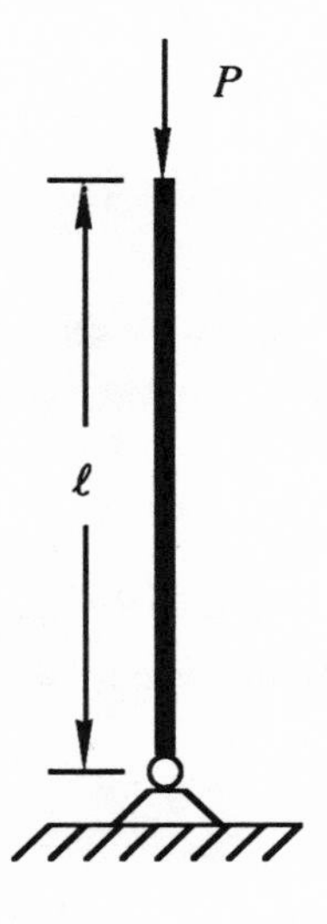

Figure P6.3. Column subjected to load P.

the buckling load P, and also determine the deflection after buckling. The bending rigidity of the column is D.

Hint: The potential energy of the system after deflection y is

$$I[y] = \frac{D}{2}\int_0^{\ell}(y'')^2\,dx - \frac{P}{2}\int_0^{\ell} y'^2\,dx.$$

6.4 The column in Prob. (6.3) has the nonuniform section $D = D_0 x/\ell$ and load P inclines ϕ after buckling (see Fig. P6.4). Find the buckling load P.

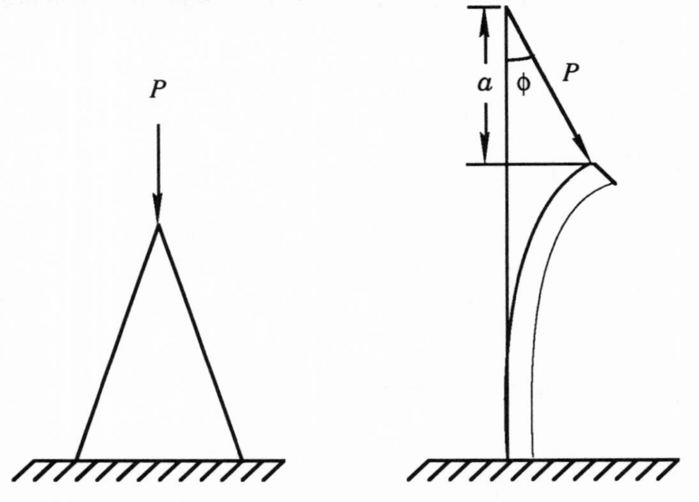

Figure P6.4. Column with nonuniform section under inclined force P.

6.5 Find a system of differential equations for the **geodesics** (shortest distance) on a surface $G(x, y, z) = 0$ passing between two points, (x_0, y_0, z_0) and (x_1, y_1, z_1).

6.6 Progressive gravity plane waves in liquid are confined in a channel, as shown in Fig. P6.6. The continuity condition of the liquid is

$$u_x + v_y = 0,$$

where u and v are velocity components of the liquid. The boundary conditions are

$$u = 0 \quad \text{at } x = 0, a,$$

$$v = 0 \quad \text{at } y = -b.$$

The kinetic energy is

$$T = \frac{1}{2}\int_0^a \int_{-b}^0 \rho\left(\dot{u}^2 + \dot{v}^2\right) dx\, dy.$$

The potential energy charge is

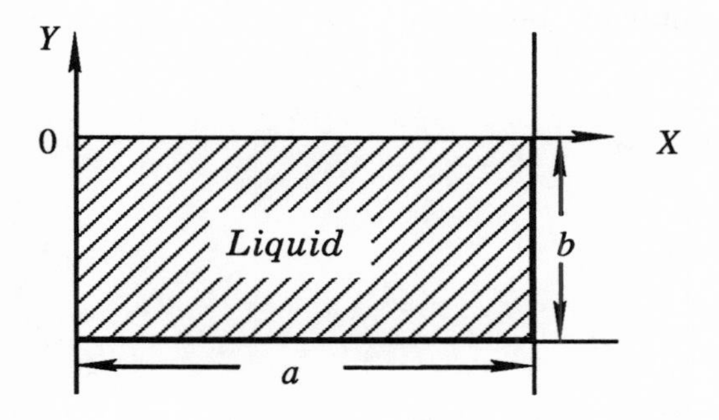

Figure P6.6. Channel with rectangular cross section.

$$U = \int_0^a \int_{-b}^0 \rho g y \, dx\, dy,$$

where ρ is the density and g is the gravitation constant. Find approximate solutions for u and v.

7

Continuity Conditions

Let us consider a problem to minimize the functional

$$I[y] = \int_{-1}^{1} x^2 (y')^2 \, dx \tag{7.1}$$

with boundary conditions $y(1) = 1$ and $y(-1) = -1$.

The Euler equation becomes

$$\left(x^2 y'\right)' = 0. \tag{7.2}$$

Then we have

$$x^2 y' = c_1, \quad y' = \frac{c_1}{x^2}, \quad y = -\frac{c_1}{x} + c_2. \tag{7.3}$$

The boundary conditions yield $c_1 = -1$ and $c_2 = 0$. The solution of (7.2) is obtained as

$$y = \frac{1}{x}. \tag{7.4}$$

However, this is not the real solution, since y is not continuous at $x = 0$. We have assumed proper continuity conditions in the fundamental lemma of calculus of variation in Chapter 2.

In order to study the required conditions for continuity, we start from the simple problem of minimizing (or maximizing) the functional

$$I[y] = \int_{x_0}^{x_1} F(x, y, y') \, dx, \tag{7.5}$$

where the values of x_0, x_1, $y(x_0)$, and $y(x_1)$ are given.

The first variation becomes

$$\delta I = \int_{x_0}^{x_1} \left(F_y - F_{y'x} - F_{y'y'} y' - F_{y'y'} y''\right) \delta y \, dx \tag{7.6}$$

after integration by parts and using $\delta y(x_1) = \delta y(x_0) = 0$. When the fundamental lemma in Chapter 2 is employed, the function $\left(F_y - F_{y'x} - F_{y'y}y' - F_{y'y'}y''\right)$ must be assumed to be piecewise continuous. Namely, we have to assume that y and y' are continuous and y'' is piecewise continuous. This requirement appears unnaturally restrictive from the point view of the variational problem; the maximum or minimum problem of the functional with the integrand $F(x,y,y')$ has a meaning even if y' is required to be only piecewise continuous and no assumptions at all regarding the second derivatives are made. It is a priori conceivable that if the conditions of admissibility are broadened in this way, one might obtain a new solution that no longer satisfies the Euler equation. But this is not the case. It is proven that the piecewise continuous function y' is actually continuous and possesses a continuous derivative if the Legendre condition

$$F_{y'y'} \neq 0 \tag{7.7}$$

is satisfied.

We start our discussion with

$$\delta I[y] = \int_{x_0}^{x_1} \left(F_y \delta y + F_{y'} \delta y'\right) dx = 0. \tag{7.8}$$

We put

$$F_{y'} = B, \quad \int_{x_0}^{x} F_y \, dx = A(x), \quad \left(F_y = A'\right). \tag{7.9}$$

Then, since $\delta y = 0$ at x_0 and x_1,

$$\delta I = \int_{x_0}^{x_1} (A' \delta y + B \delta y') dx = \int_{x_0}^{x_1} (B - A) \delta y' \, dx = 0. \tag{7.10}$$

On the other hand,

$$\int_{x_0}^{x_1} \delta y' \, dx = [\delta y]_{x_0}^{x_1} = 0. \tag{7.11}$$

Therefore, according to the **Du Bois-Reymond's theorem** (the **theorem of Haar**),

$$B = A = \text{constant} \tag{7.12}$$

or

$$F_{y'} - \int_{x_0}^{x} F_y\, dx = \text{constant} = C. \tag{7.13}$$

Since $\int_{x_0}^{x} F_y\, dx$ is differentiable, (7.13) leads to the Euler differential equation

$$\frac{d}{dx}\left(F_{y'}\right) - F_y = 0. \tag{7.14}$$

The Du-Bois-Reymond's theorem (the theorem of Haar) is as follows. If $\psi(x)$ is a piecewise continuous function and

$$\int_{x_0}^{x_1} \psi(x)\eta(x)\, dx = 0, \tag{7.15}$$

$$\int_{x_0}^{x_1} \eta(x)\, dx = 0 \tag{7.16}$$

for any piecewise continuous function $\eta(x)$ then $\psi(x)$ is constant.

Let us define a constant C as

$$\int_{x_0}^{x_1} (\psi - C)\, dx = 0. \tag{7.17}$$

Then

$$\int_{x_0}^{x_1} (\psi - C)\eta\, dx = \int_{x_0}^{x_1} \psi\eta\, dx - C\int_{x_0}^{x_1} \eta\, dx = 0. \tag{7.18}$$

Since η is an arbitrary function satisfying (7.16), it can be taken as $(\psi - C)$. Then (7.18) means

$$\int_{x_0}^{x_1} (\psi - C)^2\, dx = 0. \tag{7.19}$$

Therefore, ψ = constant C.

It can be said furthermore that y' is a continuous function if (7.7) is satisfied. This is because y' can be expressed from (7.13) as

$$y' = \phi\left(x, y, \int_{x_0}^{x} F_y \, dx, C\right), \tag{7.20}$$

and ϕ is a continuously differentiable function with respect to x.

In summary, the domain or *space* of **admissible functions** from which the argument functions can be selected in the extreme problems of (7.5) is the set of all functions that are continuous and have piecewise continuous first derivatives.

The Du Bois-Reymond's result may be extended to an integrand of the form $F(x,y,y',\ldots,y^{(n)})$. The domain of admissible functions is the set of all functions that have continuous $(n-1)$th derivatives and piecewise continuous nth derivatives.

8

Galerkin's Method

Let us consider the minimum problem of

$$I[y]=\int_0^{\ell}\left[p(y')^2+qy^2+fy\right]dx \tag{8.1}$$

with the boundary conditions

$$y(0)=y_0,\quad y(\ell)=y_1. \tag{8.2}$$

We take a complete system of functions w_1, w_2,... and put

$$y=\frac{x}{\ell}y_1+\frac{\ell-x}{\ell}y_0+a_1w_1+a_2w_2+\cdots+a_nw_n. \tag{8.3}$$

Ritz's method yields

$$\frac{\partial I}{\partial a_n}=2\int_0^{\ell}\left[py'\frac{\partial y'}{\partial a_n}+qy\frac{\partial y}{\partial a_n}+f\frac{\partial y}{\partial a_n}\right]dx=2\int_0^{\ell}(py'w_n'+qyw_n+fw_n)dx=0. \tag{8.4}$$

The above equation can be written as

$$2\int_0^{\ell}\left[-(py')'+qy+f\right]w_n\,dx=0,\quad (n=1,2,\ldots). \tag{8.5}$$

What is interesting about this result is that the integrand in the integral (8.5) is the multiplication of the Euler equation by the fundamental function. The number of equations in (8.5) is equal to the number of unknowns a_1, a_2,.... Therefore, the equations in (8.5) are sufficient to determine these unknowns.

The method (8.5) with (8.3) is called **Galerkin's method** for the differential equation $-(py')'+qy+f=0$.

If our differential equation is

$$L[u]=0 \tag{8.6}$$

and the boundary conditions are

$$u(s)=\psi(s), \quad M[u]=m(s), \tag{8.7}$$

where L and M are given differential operators, then Galerkin's method suggests an approximation method for solving the boundary -value problem.

We choose the fundamental functions to satisfy

$$w_n(s)=0, \quad M_n(s)=0, \quad (n=1,2,\ldots) \tag{8.8}$$

and

$$u_0(s)=\psi(s), \quad M[u_0]=m(s), \tag{8.9}$$

for u_0. Then Galerkin's method leads to

$$u=u_0+a_1w_1+a_2w_2+\cdots, \tag{8.10}$$

and the coefficients a_1, a_2,... are determined from

$$\iint_G L[u]w_n\,dA=0 \tag{8.11}$$

where dA is the surface element of G where the differential equation is defined.

The following is an interpretation of this method. Suppose that we have a functional

$$I[u]=\iint_G F\left(x,y,u,u_x,u_{xx},\ldots,u_y,u_{yy},\ldots,u_{xy},\ldots\right)dx\,dy, \tag{8.12}$$

and its first variation leads to

$$\delta I=\iint_G L[u]\delta u\,dx\,dy=0 \tag{8.13}$$

after the boundary terms that result from integration by parts vanish. Then Galerkin's method is a direct method for finding a stationary function corresponding to (8.12), or it is an approximation method for solving Euler's equation (8.6). This last interpretation is more important because Galerkin's method is an approximation method of boundary-value problems. The important point here is that we do not need to know the explicit form of F or I

in (8.12). It is simply sufficient to construct equation (8.13) from a given differential equation. Sometimes there is no such integrand or functional, whose Euler's equation provides a given differential equation. The equation (8.13) is sometimes called the **principle of virtual work.**

Example 8.1.
Find an approximate solution for the stress field in a rectangular plate under the applied load shown in Fig. 8.1.

Denoting Airy's stress function by ψ, the differential equation for ψ becomes

$$\frac{\partial^4\psi}{\partial x^4} + 2\frac{\partial^4\psi}{\partial x^2\partial y^2} + \frac{\partial^4\psi}{\partial y^4} = 0 \quad \text{in } G. \tag{8.14}$$

Since

$$\sigma_x = \frac{\partial^2\psi}{\partial y^2}, \quad \sigma_y = \frac{\partial^2\psi}{\partial x^2}, \quad \sigma_{xy} = -\frac{\partial^2\psi}{\partial x\,\partial y},$$

the boundary conditions are written as

$$\begin{aligned} &\psi_{yy} = s\left(1 - \frac{y^2}{b^2}\right), \quad \psi_{xy} = 0 &&\text{at } x = \pm a, \\ &\psi_{xx} = \psi_{xy} = 0 &&\text{at } y = \pm b. \end{aligned} \tag{8.15}$$

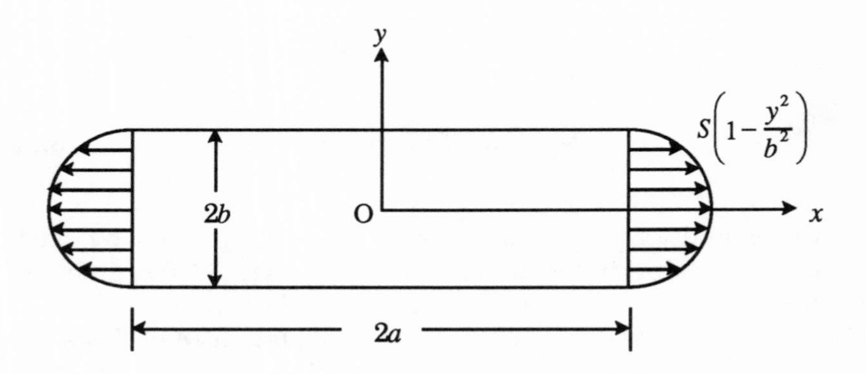

Figure 8.1. Rectangular plate under parabolic force.

We choose

$$\psi = \tfrac{1}{2} s y^2 \left(1 - \frac{y^2}{6b^2}\right) + \left(x^2 - a^2\right)\left(y^2 - b^2\right)\left(a_1 + a_2 x^2 + a_3 y^2 + \cdots\right) \tag{8.16}$$

so that it satisfies the boundary conditions (8.15). Substitute (8.16) into

$$\int_{-b}^{b}\int_{-a}^{a} \left(\nabla^4 \psi\right) \delta\psi \, dx\, dy = 0, \tag{8.17}$$

where

$$\delta\psi = \left(x^2 - a^2\right)^2 \left(y^2 - b^2\right)^2 \left(\delta a_1 + \delta a_2 x^2 + \cdots\right). \tag{8.18}$$

As a first approximation, we take only a_1 to be nonzero. Then (8.17) becomes

$$\int_{-b}^{b}\int_{-a}^{a} \left[24\left(y^2 - b^2\right)^2 a_1 + 32\left(3x^2 - a^2\right)\left(3y^2 - b^2\right) a_1 + 24\left(x^2 - a^2\right) a_1 - \frac{2s}{b^2}\right] \left(x^2 - a^2\right)^2 \left(y^2 - b^2\right)^2 dx\, dy = 0, \tag{8.19}$$

which yields

$$a_1 = \frac{s/b^2}{\frac{64}{7} b^4 + \frac{256}{49} a^2 b^2 + \frac{64}{7} a^4}, \tag{8.20}$$

$$\begin{aligned} \sigma_{xx} &= s\left(1 - \frac{y^2}{b^2}\right) + 4\left(x^2 - a^2\right)^2 \left(3y^2 - b^2\right) a_1 \\ \sigma_{yy} &= 4\left(y^2 - a^2\right)\left(3x^2 - a^2\right) a_1, \\ \sigma_{xy} &= -16\left(x^2 - a^2\right)\left(y^2 - b^2\right) xy. \end{aligned} \tag{8.21}$$

Example 8.2

Solve the equation of motion of a particle

$$m\ddot{x} + v\dot{x} + kx = 0 \tag{8.22}$$

with the initial conditions

$$\begin{aligned} x &= 0 \quad \text{at } t = 0, \\ \dot{x} &= v_0 \quad \text{at } t = 0. \end{aligned} \tag{8.23}$$

It is difficult to use Ritz's method because there is no functional form $I[x]=\int_0^{t_1} F(t,x,\dot{x})\,dt$. This difficulty is overcome, however, when Galerkin's method is employed. This method is used for finding x to satisfy

$$\int_0^{t_1} (m\ddot{x} + \nu\dot{x} + kx)\delta x\, dt = 0, \tag{8.24}$$

where x is subjected to the conditions described in (8.23).

Assume

$$x(t) = v_0 e^{-t} \sin t + a_1 e^{-t}(\cos t - 1). \tag{8.25}$$

It is obvious that this trial function satisfies the initial conditions (8.23) for any arbitrary value of a_1. Then we have

$$\delta x = e^{-t}(\cos t - 1)\delta a_1. \tag{8.26}$$

When (8.25) and (8.26) are substituted into (8.24) and the integrals are carried out by taking $t_1 = \infty$, we have

$$m(-2v_0 + 2a_1) - 2\nu v_0 + k(3v_0 - 3a_1) = 0. \tag{8.27}$$

Then

$$a_1 = \frac{(2m + 2\nu - 3k)v_0}{2m - 3k}. \tag{8.28}$$

A numerical calculation is shown in Fig. 8.2 for a special case, $m = 1$, $\nu = 1$, $k = 1$, and $v_0 = 1$. The exact solution is obtained by solving the differential equation (8.22) with initial conditions (8.23) is

$$x = e^{-\frac{1}{2}} \tfrac{2}{\sqrt{3}} \sin \tfrac{\sqrt{3}}{2} t. \tag{8.29}$$

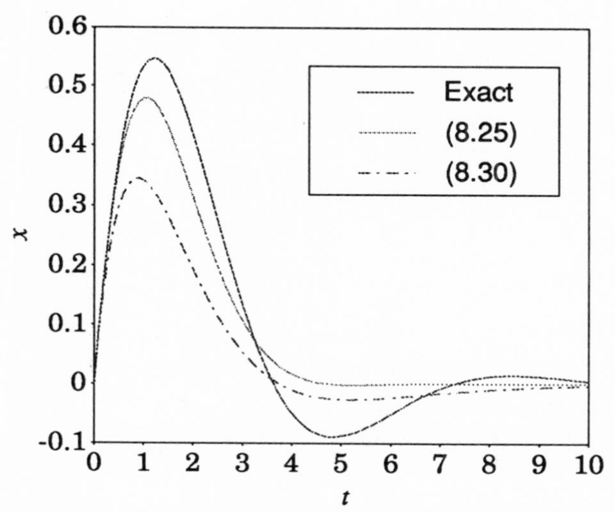

Figure 8.2. Approximate solutions for (8.22).

It is plotted in Fig. 8.2 for comparison.

Another approximation is taken as

$$x(t) = v_0 e^{-t}\left(1 - a_1 t^2\right)t, \tag{8.30}$$

which satisfies (8.23). A similar calculation leads to $a_1 = (2k - v) / (15k - m)$. The result is also plotted in Fig. 8.2.

Example 8.3.

Force f is acting on a beam with bending rigidity D. The end of the beam $x = 0$ is clamped and the other end $x = \ell$ is simply supported.

When Ritz's method is used, the functional,

$$I[y] = \frac{D}{2}\int_0^\ell (y'')^2\,dx - \int_0^\ell f y\,dx \tag{8.31}$$

is to be minimized with the boundary conditions

$$\begin{aligned} &y = 0, \quad y' = 0 && \text{at } x = 0, \\ &y = 0 && \text{at } x = \ell. \end{aligned} \tag{8.32}$$

Trial functions do not need to satisfy the natural boundary condition

$$y'' = 0 \text{ at } x = \ell. \tag{8.33}$$

When Galerkin's method is employed, the trial functions y must satisfy all the boundary conditions (8.32) and (8.33).

The equation to solve is

$$\int_0^{\ell} (Dy'''' - f)\delta y \, dx = 0. \tag{8.34}$$

Galerkin's Method by Mathematica

As Mathematica proves to be an effective tool for Ritz's method in Chapter 3, so it is for Galerkin's method. The use of Mathematica for Galerkin's method is illustrated in the Appendix.

Galerkin-Kantorovich Method

Kantorovich's method, described in Chapter 3, can be adapted to Galerkin's method as well. Consider a Poisson's equation

$$\frac{\partial^2 u}{\partial x^2} + \frac{\partial^2 u}{\partial y^2} = \sin x \tag{8.35}$$

with the boundary conditions

$$u(0, y) = u(\pi, y) = 0, \quad \left.\frac{\partial u}{\partial y}\right|_{y=-\frac{\pi}{2}} = \left.\frac{\partial u}{\partial y}\right|_{y=\frac{\pi}{2}} = x(x - \pi).$$

Assume the trial function

$$u(x, y) = x(x - \pi)f(y) \tag{8.36}$$

where $f(y)$ is an undetermined function of y. Note that the trial function satisfies the boundary conditions in the x direction. Substitution of (8.36) into (8.35) yields

$$2f + x(x - \pi)\frac{\partial^2 f}{\partial y^2} - \sin x = 0. \tag{8.37}$$

The weight functions w_n in (8.5) correspond to $x(x - \pi)$ in (8.36); therefore, when Galerkin's method is applied to the above equation, an ordinary differential equation is formed as shown below.

$$\int_0^{\pi}\left[2f(y)+x(x-\pi)\frac{\partial^2 f}{\partial y^2}-\sin x\right]x(x-\pi)dx=\frac{\partial^2 f}{\partial y^2}-\frac{10}{\pi}f+\frac{120}{\pi^5}=0 \qquad (8.38)$$

The solution of (8.38) is

$$f(y)=c_1\cosh\tfrac{\sqrt{10}}{\pi}y+c_2\sinh\tfrac{\sqrt{10}}{\pi}y+\tfrac{12}{\pi^3} \qquad (8.39)$$

where c_1 and c_2 are undetermined constants. The trial function now has the form

$$u(x,y)=x(x-\pi)\left(c_1\cosh\tfrac{\sqrt{10}}{\pi}y+c_2\sinh\tfrac{\sqrt{10}}{\pi}y+\tfrac{12}{\pi^3}\right). \qquad (8.40)$$

The undetermined constant C_1 and C_2 are computed from the remaining boundary conditions.

$$\begin{cases} x(x-\pi)\left(-c_1\tfrac{\sqrt{10}}{\pi}\sinh\tfrac{\sqrt{10}}{2}+c_2\tfrac{\sqrt{10}}{\pi}\cosh\tfrac{\sqrt{10}}{2}\right)=x(x-\pi), \\ x(x-\pi)\left(c_1\tfrac{\sqrt{10}}{\pi}\sinh\tfrac{\sqrt{10}}{2}+c_2\tfrac{\sqrt{10}}{\pi}\cosh\tfrac{\sqrt{10}}{2}\right)=x(x-\pi). \end{cases} \qquad (8.41)$$

(8.41) implies

$$\begin{cases} -c_1\tfrac{\sqrt{10}}{\pi}\sinh\tfrac{\sqrt{10}}{2}+c_2\tfrac{\sqrt{10}}{\pi}\cosh\tfrac{\sqrt{10}}{2}=1, \\ c_1\tfrac{\sqrt{10}}{\pi}\sinh\tfrac{\sqrt{10}}{2}+c_2\tfrac{\sqrt{10}}{\pi}\cosh\tfrac{\sqrt{10}}{2}=1. \end{cases} \qquad (8.42)$$

The undetermined constants are

$$c_1=0,\quad c_2=\tfrac{\pi}{\sqrt{10}}\left(\cosh\tfrac{\sqrt{10}}{2}\right)^{-1}. \qquad (8.43)$$

Therefore, the Kantorovich-Galerkin solution becomes

$$x(x-\pi)\left[\tfrac{\pi}{\sqrt{10}}\left(\cosh\tfrac{\sqrt{10}}{2}\right)^{-1}\sinh\tfrac{\sqrt{10}}{\pi}y+\tfrac{12}{\pi^3}\right]. \qquad (8.44)$$

More About Galerkin's Method

In this section, a more general procedure to apply Galerkin's method is considered. The Galerkin-type formulation, such as (8.5) is based on the assumption that the trial function satisfies both the essential and natural boundary conditions. However, in some situations to determine such a trial function may be difficult. Convenience is apparent if Galerkin's method can be applied without forcing the trial function to satisfy the boundary conditions initially.

Suppose we have a differential equation

$$-(py')' + qy + f = 0 \tag{8.45.1}$$

with the boundary conditions

$$y(0) = y_0, \quad y(\ell) = y_\ell. \tag{8.45.2}$$

We assume a trial function $\bar{y}(x,c)$ and substitute into (8.45.1) and (8.45.2). Since it is rare that we choose the exact solution as the trial function, we have

$$\begin{aligned} -(p\bar{y}')' + q\bar{y} + f &= \mathrm{E}(x,c), \\ \bar{y}(0) - y_0 &= \varepsilon_1(x,c), \\ \bar{y}(\ell) - y_\ell &= \varepsilon_2(x,c) \end{aligned} \tag{8.46}$$

where $\mathrm{E}(x,c)$, $\varepsilon_1(x,c)$, and $\varepsilon_2(x,c)$ are errors due to using the trial function. Applying Galerkin's method to (8.46) yields

$$\varepsilon_1(x,c)w_i(0) + \varepsilon_2(x,c)w_i(\ell) + \int_0^\ell \mathrm{E}(x,c)w_i\,dx = 0. \tag{8.47}$$

We have been using trial functions that satisfy the boundary conditions, so that ε_1 and ε_2 vanish. However, we still can use (8.47) to solve the same problem without initially constructing a trial function that does not satisfy the given boundary conditions.

Example 8.4.
Repeat the problem in (3.28).

Assume the trial function

$$u \cong \bar{u} = \sum_{k=0}^{6} c_k x^k. \tag{8.48}$$

Notice that this trial function does not satisfy the boundary conditions automatically. From this trial function, the weight functions are

$$\begin{aligned} w_i(x) &= \{1 \quad x \quad x^2 \quad x^3 \quad x^4 \quad x^5 \quad x^6\}, \\ w_i(0) &= \{1 \quad 0 \quad 0 \quad 0 \quad 0 \quad 0 \quad 0\}, \\ w_i(1) &= \{1 \quad 1 \quad 1 \quad 1 \quad 1 \quad 1 \quad 1\}. \end{aligned} \tag{8.49}$$

When (4.38), (8.48), and (8.49) are substituted into (8.47), we have

$$\begin{Bmatrix} 1 \\ 0 \\ 0 \\ 0 \\ 0 \\ 0 \\ 0 \end{Bmatrix} \bar{u}(0) + \begin{Bmatrix} 1 \\ 1 \\ 1 \\ 1 \\ 1 \\ 1 \\ 1 \end{Bmatrix} \bar{u}(\ell) + \int_0^1 \left[\bar{u}'' + \left(1 + x^2\right)\bar{u} + 1\right] \begin{Bmatrix} 1 \\ x \\ x^2 \\ x^3 \\ x^4 \\ x^5 \\ x^6 \end{Bmatrix} dx = 0. \tag{8.50}$$

When this equation is solved for the constants c_k and substituted back into (8.48), the solution is

$$\begin{aligned} \bar{u} = {} & \frac{5711294844996}{565794438970473775} + \frac{1450427036934197208}{2602654419264179365} x \\ & + \frac{1301825457152803176}{2602654419264179365} x^2 - \frac{239091470488891152}{2602654419264179365} x^3 \\ & + \frac{20601334100242038}{520530883852835873} x^4 - \frac{28982872858664181 6}{13013272096320896825} x^5 \\ & + \frac{45372755464524774}{2602654419264179365} x^6. \end{aligned} \tag{8.51}$$

Notice that (8.51) does not satisfy the boundary conditions exactly. However, the difference between this solution and (3.31) is very small. The conclusion here is that even though an approximate solution can be obtained without constructing a trial function that does not satisfy the boundary condition automatically, a very close solution may be obtained. However, in order to improve the solution, choosing a trial function that does satisfy the boundary conditions is more preferable.

Problems

8.1 Find an approximate solution using Galerkin's method for the Bessel differential equation

$$x^2y'' + xy' + (x^2 - 1)y = 0, \quad 1 < x < 2,$$
$$y(1) = 1, \quad y(2) = 2.$$

8.2 Find an approximate solution using Galerkin's method for the differential equation

$$[(x+2)y'']'' + y - 3x = 0, \quad 0 < x < 1$$

with the boundary conditions

$$(x+2)y'' = 0, \quad \{(x+2)y''\}' = 0 \qquad \text{at } x = 0,$$
$$y = 0, \quad y' = 0 \qquad \text{at } x = 1.$$

Hint:

$$y = a_1w_1 + a_2w_2 + \cdots,$$
$$w_1 = (x-1)^2(x^2 + 2x + 3),$$
$$w_2 = (x-1)^3(3x^2 + 4x + 3).$$

8.3 Solve the boundary-value problem

$$x^2y'' + y - 3x^2 = 0, \quad 0 < x < 1$$

with the boundary conditions $y(0) = 0$, $y(1) = 1$.

8.4 Find the approximate solution using Galerkin's method for the differential equation

$$x^2y + x(y' + 1) + (x^2 - 1)(y + x) = 0, \quad 1 < x < 2$$
$$y(1) = y(2) = 0.$$

8.5 A simply supported plate is subjected to the x direction force

$$f(y) = N_x - N_0\left[1 - \frac{ay}{b}\right]$$

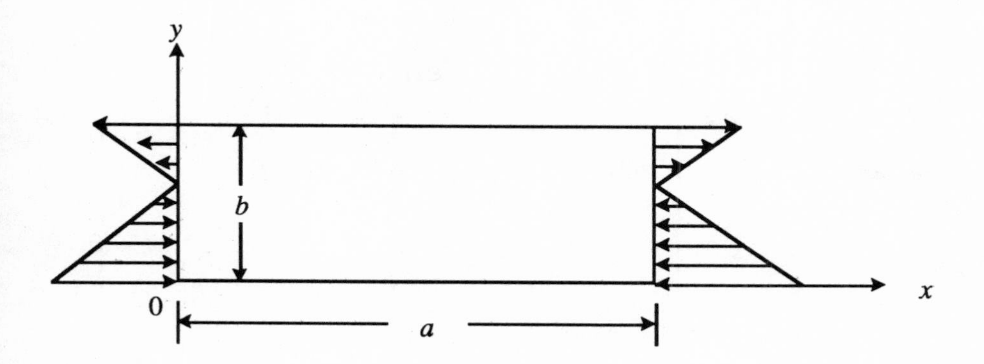

Figure P8.5. Plate buckling.

where $N_0 > 0$ and $2 \geq \alpha \geq 1$ (see Fig. P8.5). Find the buckling load N_0, where the deflection of the plate w satisfies the following conditions:

$$\frac{\partial^4 w}{\partial x^4} + 2\frac{\partial^4 w}{\partial x^2 \partial y^2} + \frac{\partial^4 w}{\partial y^4} - \frac{N_x}{P}\frac{\partial^2 w}{\partial x^2} = 0,$$

$$w = \frac{\partial^2 w}{\partial x^2} = 0 \qquad \text{at } x = 0, a,$$

$$w = \frac{\partial^2 w}{\partial y^2} = 0 \qquad \text{at } y = 0, b.$$

8.6 Consider the bending beam problem in Example 8.3. Solve the problem by Ritz's method, Galerkin's method, and by the Euler equation. Compare the two approximations with the exact solution obtained by the Euler equation.

9

Minimizing Sequence

Let us consider a minimum problem of the functional

$$I[y] = \int_{x_0}^{x_1} F(x, y, y')\,dx \tag{9.1}$$

with the boundary conditions

$$y(x_0) = y_0, \quad y(x_1) = y_1. \tag{9.2}$$

For a set of values of I taken by all admissible comparison functions, the greatest lower bound is denoted by d. Then we can choose a set of comparison functions ψ_1, ψ_2,... such that

$$\lim_{n\to\infty} I[\psi_n] = d. \tag{9.3}$$

The set of functions (ψ_k) is called a **minimizing sequence.** Direct methods such as Ritz's method or Galerkin's method are for finding the stationary function u as the limit of ψ_n,

$$u = \lim_{n\to\infty} \psi_n. \tag{9.4}$$

In these direct methods, we assume that u is also an admissible function for the problem and

$$I[u] = d. \tag{9.5}$$

This assumption, however, is not guaranteed. Unfortunately, it sometimes fails.

Example 9.1.
Consider the minimum surface in a three-dimensional space bounded by a circle with the radius 1 on the xy plane. The surface is expressed as

$$I[z] = \iint_G \sqrt{1 + z_x^2 + z_y^2}\, dx\, dy. \tag{9.6}$$

From geometrical intuition, the stationary function is $z = 0$. The minimum surface is the circular plane G, and the minimum value is π. A minimizing sequence $\psi_1, \psi_2, \ldots$ is taken, as shown in Fig. 9.1. ψ_k is the surface of a circular cone with radius ε_k and height 1 for $\sqrt{x^2 + y^2} \leq \varepsilon_k$, and it is zero for $\sqrt{x^2 + y^2} > \varepsilon_k$, where $\varepsilon_1 > \varepsilon_2 > \varepsilon_3 > \cdots$. These functions are continuous and have piecewise continuous derivatives and therefore they are admissible comparison functions. Furthermore,

$$\lim_{n \to \infty} I[\psi_n] = \pi.$$

However, $\lim_{n \to \infty} \psi_n$ does not converge to the stationary function $z = 0$.

Example 9.2.
Minimize the functional,

$$I[y] = \int_0^{\ell} (y')^2\, dx \tag{9.7}$$

with the boundary conditions $y = 0$ at $x = 0$ and 1, where y' is piecewise continuous. The first variation of I leads to

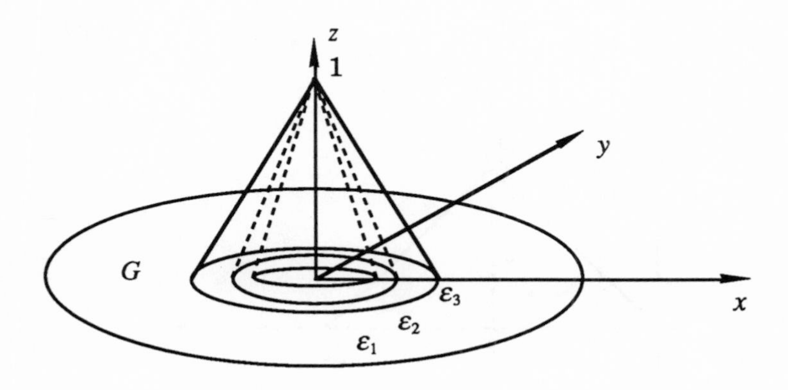

Figure 9.1. Minimizing sequence expressed by cones do not converge to the stationary function.

$$\delta I = 2\int_0^\ell y'\delta y'\,dx = -2\int_0^\ell y''\delta y\,dx = 0. \tag{9.8}$$

The Euler equation is

$$y'' = 0 \tag{9.9}$$

and the solution is $y = 0$.

Consider the minimizing sequence

$$\begin{aligned} y_n &= x && \text{for } x < \varepsilon_n, \\ y_n &= 2\varepsilon_n - x && \text{for } \varepsilon_n < x < 2\varepsilon_n, \\ y_n &= 0 && \text{for } 2\varepsilon_n < x \end{aligned} \tag{9.10}$$

where $\varepsilon_n \to 0$ for $n \to \infty$. The sequence is shown in Fig. 9.2. It is obvious that

$$\lim_{n\to\infty} I[y_n] = 0. \tag{9.11}$$

However,

$$\lim_{n\to\infty} y_n \neq y \tag{9.12}$$

where y is the stationary function. The reason is that

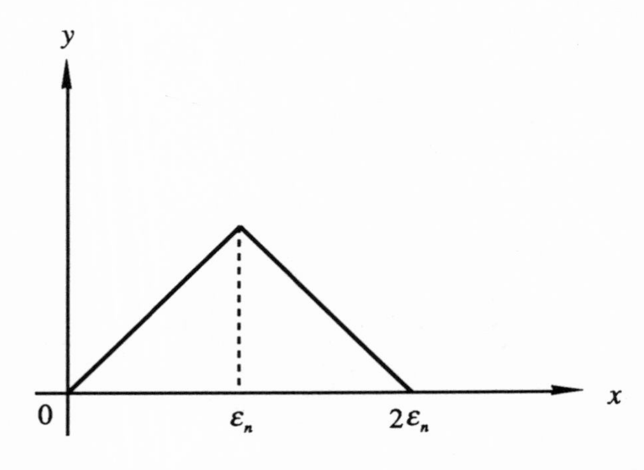

Figure 9.2. Minimizing sequences do not converge to the solution (stationary solution).

$$\lim_{n\to\infty} y_n' = 1 \qquad \text{for } x < \varepsilon_n,$$
$$\lim_{n\to\infty} y_n' = -1 \qquad \text{for } \varepsilon_n < x < 2\varepsilon_n \tag{9.13}$$

for any n. The stationary function y has zero derivative.

Convergency

Let us consider the Dirichlet integral,

$$I[\varphi] = \iint_G \left(\varphi_x^2 + \varphi_y^2\right) dx\, dy \tag{9.14}$$

where φ_x and φ_y are piecewise continuous.

Minimize (9.14) with the boundary condition at $r = 1$,

$$\varphi = \rho(\theta) = \sum_{m=1}^{\infty} \frac{1}{m^2} \cos(m!\theta), \tag{9.15}$$

where G is a circular domain $x^2 + y^2 = 1$ and (r, θ) is the polar coordinate. The integral (9.14) is written as

$$I[\varphi] = \int_0^{2\pi}\int_0^{1} \left(\varphi_r^2 + \frac{\varphi_\theta^2}{r^2}\right) r\, dr\, d\theta. \tag{9.16}$$

Therefore,

$$\begin{aligned}
\delta I &= 2\int_0^{2\pi}\int_0^1 \left(\varphi_r r \delta\varphi_r + \frac{\varphi_\theta \delta\varphi_\theta}{r}\right) dr\, d\theta \\
&= 2\left\{\int_0^{2\pi} \left[\varphi_r r \delta\varphi\right]_0^1 d\theta + \int_0^1 \left[\frac{\varphi_\theta}{r}\delta\varphi\right]_0^{2\pi} dr \right. \\
&\quad \left. - \int_0^{2\pi}\int_0^1 \frac{\partial}{\partial r}(r\varphi_r)\delta\varphi\, dr\, d\theta - \int_0^{2\pi}\int_0^1 \frac{\partial}{\partial\theta}\left(\frac{\varphi_\theta}{r}\right)\delta\varphi\, dr\, d\theta \right\} \\
&= -2\int_0^{2\pi}\int_0^1 \left[\frac{\partial}{\partial r}(r\varphi_r) + \frac{\partial}{\partial\theta}\left(\frac{\varphi_\theta}{r}\right)\right]\delta\varphi\, dr\, d\theta.
\end{aligned} \tag{9.17}$$

The Euler equation is

$$\frac{\partial}{\partial r}(r\varphi_r) + \frac{\partial}{\partial\theta}\left(\frac{\varphi_\theta}{r}\right) = 0. \tag{9.18}$$

Assuming

$$\varphi = \sum_{n=1}^{\infty} a_n r^n \cos n\theta \tag{9.19}$$

and substituting it into (9.18), we have

$$\sum_{n=1}^{\infty} a_n n^2 r^{n-1} \cos n\theta - \sum_{n=1}^{\infty} a_n n^2 r^{n-1} \cos n\theta = 0. \tag{9.20}$$

Therefore, (9.19) seems to be the solution of (9.18) if it satisfies the boundary condition (9.15). The boundary condition (9.15) leads to

$$n = m!, \quad a_n = \frac{1}{m^2}. \tag{9.21}$$

Therefore, we have

$$\varphi = \sum_{m=1}^{\infty} \frac{r^{m!}}{m^2} \cos(m!\theta). \tag{9.22}$$

Unfortunately, this is not the solution for (9.18) with (9.15) because φ_r and φ_θ, which are derived from (9.22), do not converge. As a matter of fact, the Dirichlet integral (9.14) with (9.22) becomes

$$I[\varphi] = \pi \sum_{m=1}^{\infty} \frac{m!}{m^4}. \tag{9.23}$$

This series does not converge and I becomes unbounded. The Dirichlet problem has a solution only when the boundary value $\rho(\theta)$ is extended to G and

$$D[\rho] = \iint_G \left(\rho_x^2 + \rho_y^2\right) dx\, dy \tag{9.24}$$

is bounded. This is called **Dirichlet's principle.**

10

Transformations in Variational Problems

The Lagrange multiplier method leads to several transformations, which are important both theoretically and practically. For a given minimum problem with minimum d, we shall often be able to find an equivalent maximum problem with the same value d as maximum. This is a useful tool for the practical problem of bounding d from above and below. The upper and lower bounds of the rigidity of a bar with an arbitrary cross section have been found by E. Trefftz by the use of this transformation method (E. Trefftz, Ein Gegenstck zum Ritzschen Verfahren, *Int. Kongr. Technische Mechanik,* Zürich, 2, 131, 1927).

The following principles are self-evident.

General Principle 1

Suppose that a function $f(x_1,x_2,\ldots,x_n)$ has a stationary value at $x_i=\zeta_i$ $(i=1,2,\ldots,n)$. If $f(x_1,x_2,\ldots,x_n)$ does not satisfy the relation $g(x_1,x_2,\ldots,x_n)$ at $x_i=\zeta_i$ $(i=1,2,\ldots,n)$, the stationary value of f and its stationary point do not change even if the condition $g(x_1,x_2,\ldots,x_n)=0$ is added as a constraint condition from the beginning for this stationary problem.

General Principle 2

Suppose that a function $f(x_1,x_2,\ldots,x_n)$ takes a minimum value at $x_i=\zeta_i$ $(i=1,2,\ldots,n)$. If $f(x_1,x_2,\ldots,x_n)$ does not satisfy the relation $g(x_1,x_2,\ldots,x_n)=0$ at $x_i=\zeta_i$ $(i=1,2,\ldots,n)$, the minimum value of f increases when the condition $g(x_1,x_2,\ldots,x_n)=0$ is added as a constraint condition from the beginning for this minimum problem. If the problem is a maximum problem, the maximum value decreases.

These general principles are equally applicable to variational problems of functionals.

Let us consider the following problems.

Problem I

Minimize a function $f(x,y)$ under a constraint condition $g(x,y)=0$, where the necessary conditions for the existence of the minimum value, d, are assumed to be satisfied.

Problem I is equivalent to the next problem.

Problem II

Extremize the function

$$F(x,y;\lambda)=f(x,y)+\lambda g(x,y), \tag{10.1}$$

where λ is the Lagrange multiplier.

It should be emphasized here that Problem II is generally not a minimum problem even though Problem I is a minimum problem with the minimum value d. But the stationary value of $f(x,y;\lambda)$ is equal to d.

The natural conditions for Problem II are

$$\begin{aligned} &f_x+\lambda g_x=0,\\ &f_y+\lambda g_y=0,\\ &g=0. \end{aligned} \tag{10.2}$$

The last condition in (10.2) is obtained by $\partial F/\partial\lambda=0$. At the stationary point, the conditions in (10.2) are satisfied. Therefore, according to the above general principles, the addition of the conditions in (10.2) to Problem II does not influence the extremum value of F (which is equal to the minimum value of f) or the stationary point. Thus, we have another equivalent problem.

Problem III

Extremize the function

$$F(x,y;\lambda)=f(x,y)+\lambda g(x,y) \tag{10.3}$$

with the constraint conditions

$$\begin{aligned} &f_x+\lambda g_x=0,\\ &f_y+\lambda g_y=0. \end{aligned} \tag{10.4}$$

The above problem is solved as follows.

Express x and y as functions of λ by solving (10.4), substitute them into (10.3), and then extremize them with respect to λ.

Therefore, Problem III is equivalent to the following problem.

Problem I*

Extremize the function

$$\Phi(\lambda) = f\big(x(\lambda), y(\lambda)\big) + \lambda g\big(x(\lambda), y(\lambda)\big) \tag{10.5}$$

with respect to λ, where $x(\lambda)$ and $y(\lambda)$ are obtained from (10.4).

It is obvious that the stationary values of Problems I, II, III, and I* are equal to d. We can prove that Problem I* is actually a maximum problem when the following problem, which is named II*, is a minimum problem.

Problem II*

For a given value of λ, minimize the function

$$F(x,y) = f(x,y) + \lambda g(x,y). \tag{10.6}$$

Problem II and Problem II* are different because λ is a variable in Problem II, but λ is a fixed value in Problem II*. The minimum value in Problem II* depends on λ and therefore is denoted by d_λ. It holds that

$$d_\lambda \le d. \tag{10.7}$$

Because Problem II* becomes Problem I if condition $g(x,y) = 0$ is added as a constraint condition, condition $g(x,y) = 0$ is not a natural condition of Problem II*. According to General Principle 2, the addition of a constraint condition raises its minimum value if the constraint condition is not a natural condition.

Now consider all possible values of d_λ, as a function of λ. Then Problem I* is nothing more than the problem of finding the maximum value of d_λ, since the natural conditions for Problem II* become condition (10.4) and, furthermore, the stationary value of Problem III or Problem I* is equal to d.

The important point in the above discussion is that the *minimum* problem or Problem I is transformed into the *maximum* problem of Problem I*, and the minimum value d is identical to the maximum value d. This situation suggests an error estimation method. If an approximate solution in Problem I is denoted by d_1 and that in Problem I* by d_1^*, we have

$$d_1^* \le d \le d_1. \tag{10.8}$$

The above inequality provides a range of d.

If Problem I is a maximum problem, the associated (or *complementary*) problem, Problem I*, becomes a minimum problem.

We can extend the argument above to functionals.

Problem I

Minimize the functional

$$I[y] = \int_{x_0}^{x_1} F(x, y, y')\,dx \tag{10.9}$$

with the boundary conditions

$$y(x_0) - y_0 = 0, \quad y(x_1) - y_1 = 0, \tag{10.10}$$

with $y' = dy/dx$ or

$$\frac{dy}{dx} - y' = 0. \tag{10.11}$$

Assume that this problem is a minimum problem with minimum d.

Let us construct the functional $\Phi[\lambda]$ corresponding to $\Phi(\lambda)$ in (10.5). In order to do that the constraint conditions (10.10) and (10.11) are added to the functional in (10.9) by the use of Lagrange multipliers $\lambda(x)$, μ_1, and μ_0, where y and y' are treated as independent arguments. The new functional becomes

$$J[y, y', \lambda, \mu_0, \mu_1] = \int_{x_0}^{x_1} \left[F(x, y, y') + \lambda\left(\frac{dy}{dx} - y'\right)\right] dx$$
$$+ \mu_1[y(x_1) - y_1] - \mu_0[y(x_0) - y_0]. \tag{10.12}$$

The stationary conditions for this functional are

$$F_y - \frac{d\lambda}{dx} = 0, \quad F_{y'} - \lambda = 0, \quad \lambda(x_0) + \mu_0 = 0, \quad \lambda(x_1) + \mu_1 = 0, \tag{10.13}$$

$$\frac{dy}{dx} - y' = 0, \quad y(x_0) - y_0 = 0, \quad y(x_1) - y_1 = 0. \tag{10.14}$$

We can construct the new functional $\Phi[\lambda]$ from (10.12) by using conditions (10.13) and construct the following maximum problem.

Problem I*

Maximize the functional

$$\begin{aligned}\Phi[\lambda] = \int_{x_0}^{x_1} \Bigg[& f\left(x, y\left(x, \lambda, \frac{d\lambda}{dx}\right), y'\left(x, \lambda, \frac{d\lambda}{dx}\right)\right) \\ & - \frac{d\lambda}{dx} y\left(x, \lambda, \frac{d\lambda}{dx}\right) - \lambda y'\left(x, \lambda, \frac{d\lambda}{dx}\right)\Bigg] dx \\ & - \lambda(x_0) y_0 + \lambda(x_1) y_1,\end{aligned} \tag{10.15}$$

where y and y' are functions of x, λ, and $d\lambda/dx$, which are functions obtained by solving (10.13).

The transformation of Problem I into Problem I* is called the **Legendre transformation**, the **involutory transformation,** or the **Friedrich transformation**.

Let an approximate solution of Problem I be denoted by d_1, and that of Problem I* by d_1^*. Then the true minimum value of I, which is d, is bounded by

$$d_1^* \le d \le d_1. \tag{10.16}$$

When λ is fixed in (10.12), the minimum of J which is d_λ depends on λ and the inequality (10.7) holds. Therefore, the stationary value of $\Phi[\lambda]$, d, is the maximum of d_λ, with respect to λ. This argument leads to (10.16).

Example 10.1

Find upper and lower bounds for the minimum value of the functional

$$I[u] = \frac{1}{2} \iint_G \left[\left(\frac{\partial u}{\partial x}\right)^2 + \left(\frac{\partial u}{\partial y}\right)^2\right] dx\,dy - \iint_G fu\,dx\,dy, \tag{10.17}$$

where f is a given constant. The domain G is a square as shown in Fig. 10.1. The boundary condition is

$$u = 0 \quad \text{on } \Gamma, \tag{10.18}$$

where Γ is the boundary of G.

An upper bound is obtained by substituting any admissible function u into (10.17).

Take

$$\begin{aligned} u &= A(a^2 - r^2), \qquad && r \leq a, \\ u &= 0, && r \geq a. \end{aligned} \tag{10.19}$$

The function defined above is admissible because it satisfies the continuity condition and the boundary conditions (10.18). When it is substituted into (10.17), we have

$$I(A) = \tfrac{1}{2}\int_0^a 4A^2r^2 2\pi r\,dr - f\int_0^a A(a^2 - r^2)2\pi r\,dr. \tag{10.20}$$

Condition $\partial I/\partial A = 0$ leads to $A = f/4$. The corresponding value of I is obtained as $-f^2\pi a^4/16$ which is an upper bound of the minimum value, d, of I.

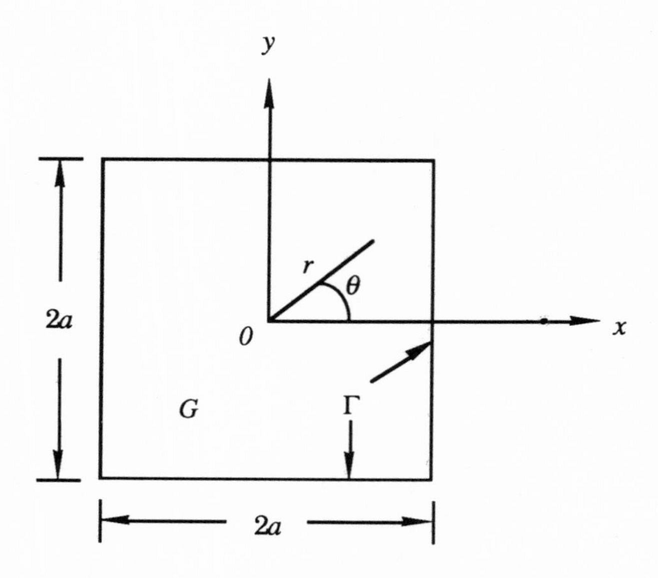

Figure 10.1. Square domain G with boundary Γ.

In order to have a lower bound of d, the Friedrich transformation is applied to (10.17).

Using the Lagrange multipliers λ_x, λ_y, and λ, the functional (10.17) is transformed into

$$I = \iint_G \left[\tfrac{1}{2}\left(u_x^2 + u_y^2\right) - \lambda_x \left(u_x - \frac{\partial u}{\partial x} \right) - \lambda_y \left(u_y - \frac{\partial u}{\partial y} \right) \right] dx\, dy$$

$$- \iint_G fu\, dx\, dy - \iint_\Gamma \lambda u\, ds. \quad (10.21)$$

The natural conditions are obtained as

$$\left. \begin{aligned} & u_x - \lambda_x = 0, \quad u_y - \lambda_y = 0, \\ & \frac{\partial \lambda_x}{\partial x} + \frac{\partial \lambda_y}{\partial y} + f = 0 \end{aligned} \right\} \text{ in } G, \quad (10.22)$$

$$\lambda_x n_x + \lambda_y n_y - \lambda = 0 \quad \text{on } \Gamma, \quad (10.23)$$

and

$$u_x = \frac{\partial u}{\partial x}, \quad u_y = \frac{\partial u}{\partial y}. \quad (10.24)$$

When conditions (10.22) and (10.23) are used, the functional (10.21) is transformed into

$$I^*[\lambda_x, \lambda_y] = -\tfrac{1}{2} \iint_G \left(\lambda_x^2 + \lambda_y^2 \right) dx\, dy. \quad (10.25)$$

Now we have the problem to maximize (10.25) with the constraint condition

$$\frac{\partial \lambda_x}{\partial x} + \frac{\partial \lambda_y}{\partial y} + f = 0 \quad \text{in } G. \quad (10.26)$$

We choose

$$\begin{aligned} \lambda_x &= Ax \\ \lambda_y &= By. \end{aligned} \quad (10.27)$$

Since (10.26) must be satisfied, we have

$$A + B + f = 0. \tag{10.28}$$

Substitute

$$\begin{aligned} \lambda_x &= Ax \\ \lambda_y &= -(A + f)y \end{aligned} \tag{10.29}$$

into (10.25); then

$$I^*(A) = -\tfrac{2}{3}a^4\left[A^2 + (f + A)^2\right]. \tag{10.30}$$

The operation $\partial I^*/\partial A = 0$ leads to $A = -f/2$ and $I^* = -a^4 f^2/3$. Thus we have the bounds

$$-\frac{a^4 f^2}{3} < d < -\frac{\pi}{16}a^4 f^2, \tag{10.31}$$

or

$$-0.333a^4 f^2 < d < -0.196a^4 f^2. \tag{10.32}$$

It is physically more meaningful if the minimum value of I has a useful quantity. Let us consider the next example.

Example 10.2

Consider a condenser of a square plate with a center hole as shown in Fig. 10.2, and find out its condenser capacity C. The electrical potential u satisfies

$$\frac{\partial^2 u}{\partial x^2} + \frac{\partial^2 u}{\partial y^2} = 0 \qquad \text{in } G, \tag{10.33}$$

$$\begin{aligned} u &= 0 \qquad \text{on } \Gamma_1(x = \pm b, \quad y = \pm b), \\ u &= 1 \qquad \text{on } \Gamma_0(r = a). \end{aligned} \tag{10.34}$$

The condenser capacity $I[u]$ is

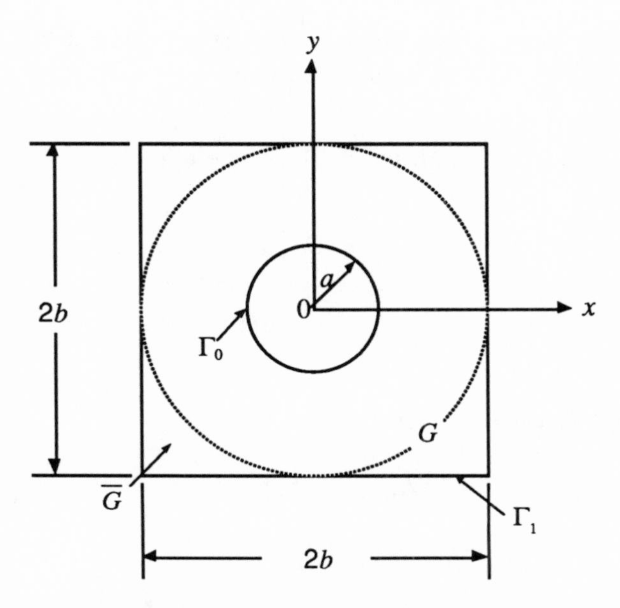

Figure 10.2. Condenser of a square plate with a center hole.

$$I[u] = \frac{1}{2\pi}\iint_G \left[\left(\frac{\partial u}{\partial x}\right)^2 + \left(\frac{\partial u}{\partial y}\right)^2\right] dx\, dy. \tag{10.35}$$

Since the potential u is obtained from the minimum problem of $I[u]$ with the boundary conditions given in (10.34), an approximation of $I[u]$ gives an upper bound for the capacity C.
Choose a trial function,

$$\begin{aligned} u &= -\log\left(\frac{r}{b}\right)\log\left(\frac{b}{a}\right) && \text{in } a \le r \le b, \\ u &= 0 && \text{elsewhere,} \end{aligned} \tag{10.36}$$

where $r = \sqrt{x^2 + y^2}$. u given by (10.36) is continuous and satisfies the boundary conditions given in (10.34). When it is substituted into (10.35), we have

$$I = \frac{1}{\log(b/a)}, \tag{10.37}$$

which gives an upper bound of C.

In order to find a lower bound of C, the Friedrich transformation is applied to $I[u]$. The new functional is constructed as

$$J[u,u_x,u_y,\lambda_x,\lambda_y,\mu_1,\mu_0] = \tfrac{1}{\pi}\iint_G \left[\tfrac{1}{2}\left(u_x^2+u_y^2\right) - \lambda_x\left(u_x - \frac{\partial u}{\partial x}\right) - \lambda_y\left(u_y - \frac{\partial u}{\partial y}\right)\right] dx\,dy - \tfrac{1}{\pi}\int_{\Gamma_1} \mu_1 u\, ds - \tfrac{1}{\pi}\int_{\Gamma_0} \mu_0 (u-1)\, ds. \tag{10.38}$$

The natural conditions obtained as

$$\left.\begin{aligned} &u_x = \lambda_x, \quad u_y = \lambda_y \\ &\frac{\partial \lambda_x}{\partial x} + \frac{\partial \lambda_y}{\partial y} = 0 \end{aligned}\right\} \quad \text{in } G,$$
$$\begin{aligned} &\lambda_x n_x + \lambda_y n_y = \mu_1 \quad &&\text{on } \Gamma_1 \\ &\lambda_x n_x + \lambda_y n_y = \mu_0 \quad &&\text{on } \Gamma_0 \end{aligned} \tag{10.39}$$

and

$$\begin{aligned} &u_x - \frac{\partial u}{\partial x} = 0, \quad u_y - \frac{\partial u}{\partial y} = 0 \quad &&\text{in } G, \\ &u = 0 &&\text{on } \Gamma_1, \\ &u = 1 &&\text{on } \Gamma_0, \end{aligned} \tag{10.40}$$

where n_x and n_y are components of the outward normal to the boundaries of G.

When λ_x and λ_y are chosen so that they satisfy conditions (10.39), the functional (10.38) is transformed into

$$I[\lambda_x,\lambda_y] = -\tfrac{1}{\pi}\iint_G \tfrac{1}{2}\left(\lambda_x^2+\lambda_y^2\right) dx\,dy + \tfrac{1}{\pi}\int_{\Gamma_0}\left(\lambda_x n_x + \lambda_y n_y\right) ds \tag{10.41}$$

Since u_x, u_y, μ_1, and μ_0 have disappeared in the last expression, the practical constraint condition for the variation for (10.41) is only

$$\frac{\partial \lambda_x}{\partial x} + \frac{\partial \lambda_y}{\partial y} = 0 \quad \text{in } G. \tag{10.42}$$

The real λ_x and λ_y which correspond to $\partial u/\partial x$ and $\partial u/\partial y$ are obtained by the maximum problem for (10.41). It is obviously a maximum problem since the first integral in (10.41) is negative definite.

Let us choose

$$\lambda_x = \frac{Ax}{r^2}, \quad \lambda_y = \frac{Ay}{r^2}, \tag{10.43}$$

so that they satisfy constraint condition (10.42). The constant A is chosen by $\partial I(A)/\partial A = 0$. Since $n_x = -x/a$ and $n_y = -y/a$ on Γ_0, when (10.43) is substituted into (10.41), we have

$$I(A) = -\tfrac{1}{2\pi} \iint_G \frac{A^2}{r^2}\, dx\, dy - 2A. \tag{10.44}$$

Then A is obtained as

$$A = \frac{-1}{1/(2\pi) \iint_G dx\, dy/r^2}. \tag{10.45}$$

The value of $I(A)$ with (10.45) gives a lower bound of C. It is

$$\begin{aligned} I(A) &= \frac{-1}{1/(2\pi) \iint_G dx\, dy/r^2} \\ &= \frac{1}{\log(b/a) + 1/(2\pi) \iint_{\overline{G}} dx\, dy/r^2}, \end{aligned} \tag{10.46}$$

where $\overline{G}$ is the domain bounded by Γ_1 and the circle with radius b. Finally, we have

$$\frac{1}{\log(b/a) + 1/(2\pi) \iint_{\overline{G}} dx\, dy/r^2} < C < \frac{1}{\log(b/a)}. \tag{10.47}$$

Example 10.3

Find the torsional rigidity of a uniform bar whose cross section is given by Fig. 10.3.

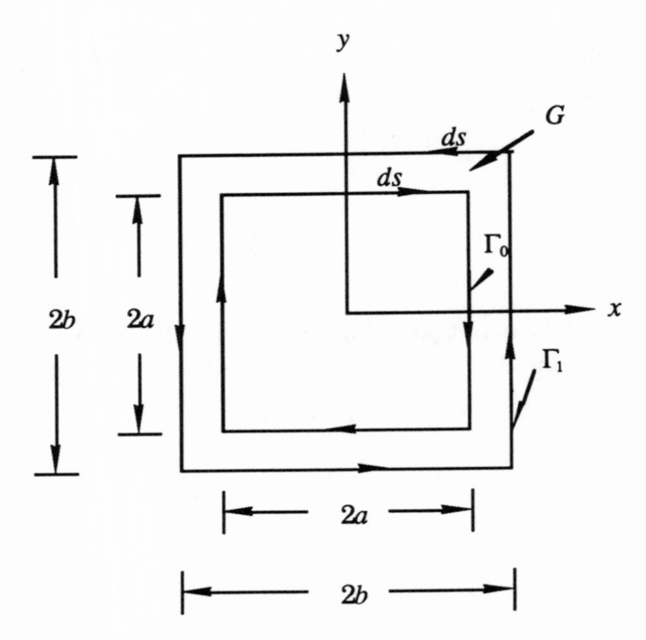

Figure 10.3. Torsional bar with cross section G bounded by Γ_0 and Γ_1.

The torsional rigidity is defined by

$$D = \frac{M}{\theta}, \tag{10.48}$$

where M is the torsional moment and θ is the twisting angle per unit length of the bar. The moment expression is

$$M = \iint_G \left(\tau_{zy} x - \tau_{zx} y\right) dx\, dy, \tag{10.49}$$

where

$$\tau_{zx} = \mu\left(\frac{\partial w}{\partial x} + \frac{\partial u}{\partial z}\right), \quad \tau_{zy} = \mu\left(\frac{\partial w}{\partial y} + \frac{\partial v}{\partial z}\right), \tag{10.50}$$

and μ is the shear modulus. As an approximation, we use the solution for a cylindrical bar with circular cross section,

$$\begin{aligned} w &= \theta \overline{w}, \\ u &= -\theta y z, \\ v &= \theta x z. \end{aligned} \tag{10.51}$$

When (10.51) is used for (10.48)–(10.50), we have

$$D = \frac{M}{\theta} = \mu \iint_G \left[\left(\frac{\partial \overline{w}}{\partial y} + x \right) x - \left(\frac{\partial \overline{w}}{\partial x} - y \right) y \right] dx\, dy$$

$$= \mu \iint_G \left(\frac{\partial \overline{w}}{\partial y} x - \frac{\partial \overline{w}}{\partial x} y \right) dx\, dy + \mu \iint_G \left(x^2 + y^2 \right) dx\, dy$$

$$= \mu \int_\Gamma \overline{w} \left(x n_y - y n_x \right) ds + \mu \iint_G \left(x^2 + y^2 \right) dx\, dy. \qquad (10.52)$$

On the boundary $\Gamma = \Gamma_0 + \Gamma_1$, the zero force traction is

$$\tau_{zx} n_x + \tau_{zy} n_y = 0 \qquad (10.53)$$

or

$$\left(\frac{\partial \overline{w}}{\partial x} - y \right) n_x + \left(\frac{\partial \overline{w}}{\partial y} + x \right) n_y = 0. \qquad (10.54)$$

Then

$$D = -\mu \int_\Gamma \overline{w} \left(\frac{\partial \overline{w}}{\partial x} n_x + \frac{\partial \overline{w}}{\partial y} n_y \right) ds + \mu \iint_G \left(x^2 + y^2 \right) dx\, dy \qquad (10.55)$$

or

$$D = -\mu \iint_G \left[\left(\frac{\partial \overline{w}}{\partial x} \right)^2 + \left(\frac{\partial \overline{w}}{\partial y} \right)^2 \right] dx\, dy + \mu \iint_G \left(x^2 + y^2 \right) dx\, dy \qquad (10.56)$$

if the condition

$$\frac{\partial^2 \overline{w}}{\partial x^2} + \frac{\partial^2 \overline{w}}{\partial y^2} = 0 \quad \text{in } G \qquad (10.57)$$

is satisfied.

It is noted that the expression for rigidity (10.56) has been derived from the original definition by assuming condition (10.54) on Γ and condition (10.57) in G.

Let us consider a maximum problem of the functional

$$I[\overline{w}] = -\mu \iint_G \left[\left(\frac{\partial \overline{w}}{\partial x} \right)^2 + \left(\frac{\partial \overline{w}}{\partial y} \right)^2 \right] dx\, dy \tag{10.58}$$

without any constraint condition or boundary condition. The first vanishing variation leads to

$$\begin{aligned}\delta I &= -2\mu \iint_G \left(\frac{\partial \overline{w}}{\partial x} \frac{\partial \delta \overline{w}}{\partial x} + \frac{\partial \overline{w}}{\partial y} \frac{\partial \delta \overline{w}}{\partial y} \right) dx\, dy + 2\mu \int_\Gamma (y n_x - x n_y) \delta \overline{w}\, ds \\ &= -2\mu \int_\Gamma \left(\frac{\partial \overline{w}}{\partial x} n_x + \frac{\partial \overline{w}}{\partial y} n_y \right) \delta \overline{w}\, ds + 2\mu \int_\Gamma (y n_x - x n_y) \delta \overline{w}\, ds \\ &\qquad + 2\mu \iint_G \left(\frac{\partial^2 \overline{w}}{\partial x^2} + \frac{\partial^2 \overline{w}}{\partial y^2} \right) \delta \overline{w}\, dx\, dy = 0. \end{aligned} \tag{10.59}$$

The natural conditions are

$$\frac{\partial^2 \overline{w}}{\partial x^2} + \frac{\partial^2 \overline{w}}{\partial y^2} = 0 \quad \text{in } G \tag{10.60}$$

and

$$\left(\frac{\partial \overline{w}}{\partial x} - y \right) n_x + \left(\frac{\partial \overline{w}}{\partial y} + x \right) n_y = 0 \quad \text{on } \Gamma. \tag{10.61}$$

Since the conditions (10.61) and (10.60) are the same as (10.54) and (10.57), we see that the expression (10.58) is related to D.

In order to find an upper bound of I, the Friedrich transformation is applied. Namely, $\partial \overline{w}/\partial x$ and $\partial \overline{w}/\partial y$ are replaced by $\overline{w}_x$ and $\overline{w}_y$, and conditions $\overline{w}_x = \partial \overline{w}/\partial x$ and $\overline{w}_y = \partial \overline{w}/\partial y$ are eliminated by the Lagrange multipliers $2\lambda_x$ and $2\lambda_y$. Then (10.58) is transformed into

$$\begin{aligned} J[\overline{w}, \overline{w}_x, \overline{w}_y, \lambda_x, \lambda_y] = -\mu \iint_G \Bigg[(\overline{w}_x^2 + \overline{w}_y^2) - 2\lambda_x \left(\overline{w}_x - \frac{\partial \overline{w}}{\partial x} \right) \\ - 2\lambda_y \left(\overline{w}_y - \frac{\partial \overline{w}}{\partial y} \right) \Bigg] dx\, dy. \end{aligned} \tag{10.62}$$

The natural conditions for the stationary value of (10.62) are obtained as

$$\left.\begin{aligned} &\overline{w}_x - \lambda_x = 0, \quad \overline{w}_y - \lambda_y = 0, \\ &\frac{\partial \lambda_x}{\partial x} + \frac{\partial \lambda_y}{\partial y} = 0 \end{aligned}\right\} \quad \text{in } G, \qquad \lambda_x n_x + \lambda_y n_y = 0 \quad \text{on } \Gamma, \tag{10.63}$$

and

$$\overline{w}_x - \frac{\partial \overline{w}}{\partial x} = 0, \quad \overline{w}_y - \frac{\partial \overline{w}}{\partial y} = 0 \quad \text{in } G. \tag{10.64}$$

Now conditions (10.63) are added as constraint conditions for the extremum problem for (10.62). Then (10.62) is transformed into

$$I^*(\lambda_x, \lambda_y) = \mu \iint_G (\lambda_x^2 + \lambda_y^2)\, dx\, dy. \tag{10.65}$$

This is a minimum problem with respect to λ_x and λ_y subjected to the constraint conditions

$$\begin{aligned} &\frac{\partial \lambda_x}{\partial x} + \frac{\partial \lambda_y}{\partial y} = 0 \quad \text{in } G, \\ &\lambda_x n_x + \lambda_y n_y = 0 \quad \text{on } \Gamma. \end{aligned} \tag{10.66}$$

A set of approximations for λ_x and λ_y will give an upper bound of I^* or I. An upper bound of I yields a lower bound of D.

Canonical Transformation

The **canonical differential equations** are alternative descriptions of Newton's equations in classical dynamics. Let us consider a mechanical system with n degree of freedom, $q_1, q_2, \ldots, q_n$. The dynamical system is determined by the minimum principle (**Hamilton's principle**) of the functional,

$$I[q_i] = \int_{t_0}^{t_1} L(t, q_i, \dot{q}_i)\, dt \tag{10.67}$$

where L is the Lagrangian,

$$L = T - U. \tag{10.68}$$

T is the kinetic energy and U is the potential energy. When q_i and $\dot{q}_i$ are treated as independent variables, (10.68) is rewritten by the use of the Lagrange multipliers p_i,

$$I[q_i, \dot{q}_i, p_i] = \int_{t_0}^{t_1} \left[L + \sum_{i=1}^{n} p_i \left(\frac{dq_i}{dt} - \dot{q}_i \right) \right] dt. \tag{10.69}$$

When the values of q_i are prescribed at $t = t_0$ and t_1 the stationary conditions for (10.69) become

$$\frac{\partial L}{\partial q_i} - \frac{dp_i}{dt} = 0, \tag{10.70}$$

$$\frac{\partial L}{\partial \dot{q}_i} - p_i = 0, \tag{10.71}$$

and

$$\frac{dq_i}{dt} - \dot{q}_i = 0. \tag{10.72}$$

If conditions (10.71) are added as the constraint conditions to (10.69), $\dot{q}_i$ can be expressed in terms of q_i and p_i. Then, (10.69) is written as

$$I[q_i, p_i] = \int_{t_0}^{t_1} \left(\sum_{i=1}^{n} p_i \frac{dq_i}{dt} - \psi \right) dt \tag{10.73}$$

where

$$\psi(t, q_i, \dot{q}_i) = \sum_{i=1}^{n} p_i \dot{q}_i - L(t, q_i, \dot{q}_i). \tag{10.74}$$

$\dot{q}_i$ in (10.74) are functions of q_i and p_i from (10.71). The stationary conditions for (10.73) are

$$\frac{\partial \psi}{\partial p_i} = \frac{dq_i}{dt}$$
$$\frac{\partial \psi}{\partial q_i} = -\frac{dp_i}{dt} \tag{10.75}$$

The function ψ is called **Hamiltonian** and the equations in (10.75) are the **canonical differential equations**. The transformation from (10.67) to (10.73) is similar to the Friedrich transformation.

Example 10.4
When $T = \frac{1}{2} a_{ij} \dot{q}_i \dot{q}_j$ and $U = \frac{1}{2} k_{ij} q_i q_j$ (the summation convention is used for repeated indices), the Hamiltonian becomes

$$\psi(q_i, p_i) = \tfrac{1}{2} a_{ij} \dot{q}_i \dot{q}_j + \tfrac{1}{2} k_{ij} q_i q_j \tag{10.76}$$

where $\dot{q}_i$ are expressed in terms of q_i and p_i by solving

$$a_{ij} \dot{q}_j - p_i = 0, \tag{10.77}$$

namely,

$$\dot{q}_i = a_{ij}^{-1} p_j \tag{10.78}$$

and

$$\psi(q_i, p_i) = \tfrac{1}{2} a_{ij}^{-1} p_i p_j + \tfrac{1}{2} k_{ij} q_i q_j \tag{10.79}$$

where a_{ij}^{-1} is the inverse matrix of a_{ij}. The canonical differential equations become

$$\frac{\partial \psi}{\partial p_i} = a_{ij}^{-1} p_j = \frac{dq_i}{dt},$$
$$\frac{\partial \psi}{\partial q_i} = k_{ij} q_j \ = -\frac{dp_i}{dt}. \tag{10.80}$$

The **conservation law of energy** $\psi = \text{const.}$ is obtained by

$$\begin{aligned}\frac{d\psi}{dt} &= \frac{\partial\psi}{\partial p_i}\frac{\partial p_i}{dt} + \frac{\partial\psi}{\partial q_i}\frac{dq_i}{dt} \\ &= \frac{dq_i}{dt}\frac{dp_i}{dt} - \frac{dp_i}{dt}\frac{dq_i}{dt} \\ &= 0.\end{aligned} \tag{10.81}$$

Problems

10.1. Find upper and lower bounds of the torsional rigidity of the bar that has the cross section in Fig. 10.3.

10.2. Find upper and lower bounds for the functional

$$I[y] = \int_0^1 \left[\tfrac{1}{2}(y' \sin \pi x)^2 - xy\right] dx$$

with the boundary conditions $y(0) = 0$ and $y(1) = 0$.

10.3 Find upper and lower bounds for the functional

$$I[y] = 2\pi \int_{x_0}^{x_1} y\sqrt{1 + (y')^2}\, dx$$

where $y(x_0) = y_0$ and $y(x_1) = y_1$.

10.4 Find upper and lower bounds for the functional

$$I[y] = \frac{T}{2}\int_0^{\ell} (y')^2\, dx - \int_0^{\ell} fy\, dx,$$

where $y = 0$ at $x = 0$ and $x = \ell$ and is a constant.

10.5 Find upper and lower bounds for the functional

$$I[\psi] = \frac{1}{2}\iint_G \left[\left(\frac{\partial\psi}{\partial x}\right)^2 + \left(\frac{\partial\psi}{\partial y}\right)^2\right] dx\, dy,$$

where G is a circular domain with radius a and $\psi = \sin\theta$ at the boundary of G. θ is the polar angle.

11

Elasticity

The Friedrich transformation can be applied to the variational problem in elasticity. We consider a three-dimensional elastic body G with boundary Γ. The body is subjected to a body force X in G and a surface traction F on a part of Γ (say Γ_1). The displacement u_i is described on the remainder of Γ (say Γ_2) (see Fig. 11.1).

The boundary conditions are

$$\sigma_{ij} n_j = F_i \qquad \text{on } \Gamma_1, \tag{11.1}$$

$$u_i = \bar{u}_i \qquad \text{on } \Gamma_2. \tag{11.2}$$

The elastic strain tensor ε_{ij} is related to the derivatives of the displacement vector u_i,

$$\varepsilon_{ij} = \tfrac{1}{2}\left(u_{i,j} + u_{j,i}\right). \tag{11.3}$$

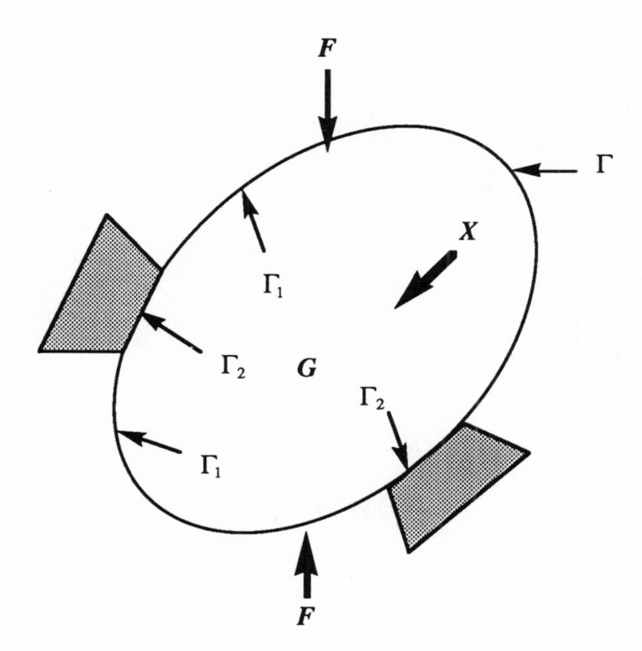

Figure 11.1. Elastic body subjected to body force X, surface force F, and clamped at boundaries Γ_2.

Hooke's law is

$$\begin{aligned}\sigma_{ij} &= C_{ijkl}\varepsilon_{kl} \\ &= C_{ijkl}u_{k,l},\end{aligned} \tag{11.4}$$

where C_{ijkl} are the elastic moduli.

The equations of equilibrium are

$$\sigma_{ij,j} + X_i = 0. \tag{11.5}$$

The boundary-value problem described by (11.1) to (11.5) is solved by the following variational problem.

We minimize the potential energy of a system,

$$I[u_i] = \tfrac{1}{2}\iiint_G \sigma_{ij}\varepsilon_{ij}\,dv - \iiint_G X_i u_i\,dv - \iint_{\Gamma_1} F_i u_i\,ds, \tag{11.6}$$

with the boundary condition

$$u_i = \bar{u}_i \qquad \text{on } \Gamma_2, \tag{11.7}$$

where σ_{ij} and ε_{ij} are functions of u_i given by (11.4) and (11.3). Since the first integral in (11.6) containing the highest derivatives of u_i is positive definite, this is a minimum problem.

It is easy to show that the natural conditions of this variational problem are (11.5) and (11.1),

$$\delta I = \tfrac{1}{2}\delta\iiint_G \sigma_{ij}\varepsilon_{ij}\,dv - \iiint_G X_i\delta u_i\,dv - \iint_{\Gamma_1} F_i\delta u_i\,ds = 0, \tag{11.8}$$

where

$$\begin{aligned}\delta\iiint_G \sigma_{ij}\varepsilon_{ij}\,dv &= \delta\iiint_G C_{ijkl}u_{k,l}u_{i,j}\,dv \\ &= \iiint_G C_{ijkl}\delta u_{k,l}u_{i,j}\,dv + \iiint_G C_{ijkl}u_{k,l}\delta u_{i,j}\,dv \\ &= \iiint_G \sigma_{kl}\delta u_{k,l}\,dv + \iiint_G \sigma_{ij}\delta u_{i,j}\,dv \\ &= 2\iiint_G \sigma_{ij}\delta u_{i,j}\,dv.\end{aligned} \tag{11.9}$$

Therefore, we have

$$\delta I = \iiint_G \sigma_{ij}\delta u_{i,j}\,dv - \iiint_G X_i\delta u_i\,dv - \iint_{\Gamma_1} F_i\delta u_i\,ds = 0. \tag{11.10}$$

After integration by parts, we have

$$\delta I = \iint_{\Gamma_1+\Gamma_2} \sigma_{ij}n_j\delta u_i\,ds - \iiint_G \left(\sigma_{ij,j} + X_i\right)\delta u_i\,dv - \iint_{\Gamma_1} F_i\delta u_i\,ds = 0, \tag{11.11}$$

where $\delta u_i = 0$ on Γ_2. The natural conditions are

$$\begin{aligned} \sigma_{ij,j} + X_i &= 0 \quad \text{in } G, \\ \sigma_{ij}n_j &= F_i \quad \text{on } \Gamma_1. \end{aligned} \tag{11.12}$$

The expression (11.11) is called the **principle of virtual work.**

Example 11.1
Find the deflection at $x = \ell/2$ of a beam with the bending rigidity EI when a concentrated load P is applied at $x = c$ (see Fig. 11.2). The potential energy of the beam is

$$I[w] = \tfrac{1}{2}\iiint \sigma_x\varepsilon_x\,dv - P(w)_{w=c} \tag{11.13}$$

with the boundary conditions

$$w = 0 \qquad \text{at } x = 0 \text{ and } x = \ell. \tag{11.14}$$

The elastic strain energy is

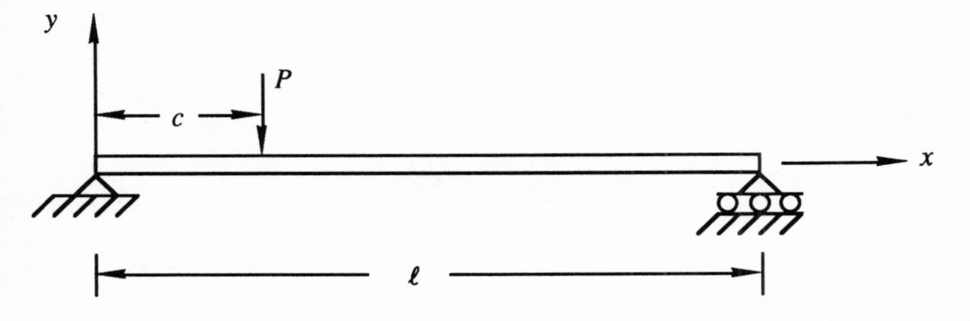

Figure 11.2. Bending beam by a concentrated force P.

$$\tfrac{1}{2}\iiint \sigma_x \varepsilon_x \, dv = \tfrac{1}{2}\iiint E\varepsilon_x^2 \, dv, \tag{11.15}$$

where E is Young's modulus. The strain is

$$\varepsilon_x = -y\frac{d^2w}{dx^2} \tag{11.16}$$

for small deflection w and

$$\iint\limits_A y^2 \, dA = I, \quad dv = dA\,dx. \tag{11.17}$$

Then (11.13) is written as

$$I[w] = \frac{EI}{2}\int_0^\ell \left(\frac{d^2w}{dx^2}\right)^2 dx - P(w)_{x=c}. \tag{11.18}$$

Assume

$$w = a_1 \sin\frac{\pi x}{\ell} + a_2 \sin\frac{2\pi x}{\ell} + \cdots \tag{11.19}$$

so that the boundary conditions (11.14) are satisfied.

When (11.19) is substituted into (11.18), we have

$$I(a_1, a_2, \ldots) = \frac{EI\pi^4}{4\ell^3}\sum_{n=1}^{\infty} n^4 a_n - P\left(a_1 \sin\frac{\pi c}{\ell} + a_2 \sin\frac{2\pi c}{\ell} + \cdots\right). \tag{11.20}$$

For the condition that the variation vanishes,

$$\frac{\partial I}{\partial a_n} = 0 \tag{11.21}$$

leads to

$$\frac{EI\pi^4}{2\ell^3} n^4 a_n - P \sin\frac{n\pi c}{\ell} = 0 \tag{11.22}$$

or

$$a_n = \frac{2P\ell^3 \sin(n\pi c/\ell)}{EI\pi^4 n^4}, \quad n = 1, 2, \ldots. \tag{11.23}$$

The deflection at $x = \ell/2$ is

$$(w)_{x=\ell/2} = \frac{P\ell^3}{48.7EI}. \tag{11.24}$$

Let us consider the Friedrich transformation of (11.6). The conditions for displacement (11.3) and (11.2) are treated by the Lagrange multipliers λ_{ij} and λ_i. Then the new functional is constructed as

$$I[u_i, \varepsilon_{ij}, \lambda_{ij}, \lambda_i] = \tfrac{1}{2}\iiint_G \sigma_{ij}\varepsilon_{ij}\, dv - \iiint_G X_i u_i\, dv - \iint_{\Gamma_1} F_i u_i\, ds$$
$$- \iiint_G \lambda_{ij}\left[\varepsilon_{ij} - \tfrac{1}{2}(u_{i,j} + u_{j,i})\right] dv - \iint_{\Gamma_2} \lambda_i (u_i - \bar{u}_i)\, ds, \tag{11.25}$$

where $\sigma_{ij} = C_{ijkl}\varepsilon_{kl}$ and $\lambda_{ij} = \lambda_{ji}$.

The stationary conditions for (11.25) with respect to the independent arguments, u_i, ε_{ij}, λ_i, are

$$\begin{aligned}
\sigma_{ij} &= \lambda_{ij} && \text{in } G, \\
\lambda_{ij,j} + X_i &= 0 && \text{in } G, \\
\lambda_{ij} n_j &= F_i && \text{on } \Gamma_1, \\
\lambda_{ij} n_j &= \lambda_i && \text{on } \Gamma_2, \\
\varepsilon_{ij} &= \tfrac{1}{2}(u_{i,j} + u_{j,i}) && \text{in } G, \\
u_i &= \bar{u}_i && \text{on } \Gamma_2.
\end{aligned} \tag{11.26}$$

The first four conditions in (11.26) are conditions for stresses. When these conditions are imposed on (11.25) as the constraint conditions for the extremum, (11.25) becomes

$$I^*[\sigma_{ij}] = -\tfrac{1}{2}\iiint_G \sigma_{ij}\varepsilon_{ij}\, dv + \iint_{\Gamma_2} \sigma_{ij} n_j \bar{u}_i\, ds, \tag{11.27}$$

where

$$\varepsilon_{ij} = C^{-1}_{ijkl}\sigma_{kl}. \tag{11.28}$$

C^{-1}_{ijkl} is the inverse matrix of C_{ijkl}. The new functional (11.27) is called the **complementary energy.** This energy takes a maximum at the equilibrium state. The constraint conditions imposed to the variations are

$$\begin{aligned} \sigma_{ij,j} + X_i = 0 \qquad &\text{in } G, \\ \sigma_{ij} n_j = F_i \qquad &\text{on } \Gamma_1. \end{aligned} \tag{11.29}$$

Example 11.2
Find the stress distribution in the rectangular bar shown in Fig. 11.3 under a given torsion M at the end.

Since $\bar{u}_i = 0$ on $\Gamma_2 (z = 0)$ the functional (11.27) becomes

$$I^*[\psi] = -\frac{\ell}{2\mu} \int_{-b}^{b} \int_{-a}^{a} \left(\tau_{zx}^2 + \tau_{zy}^2 \right) dx\, dy, \tag{11.30}$$

where μ is the shear modulus. The stress components must satisfy

$$\frac{\partial \tau_{zx}}{\partial x} + \frac{\partial \tau_{zy}}{\partial y} = 0 \qquad \text{in } G, \tag{11.31}$$

$$\int_{-b}^{b} \int_{-a}^{a} \left(\tau_{zy} x - \tau_{zy} y \right) dx\, dy = M, \tag{11.32}$$

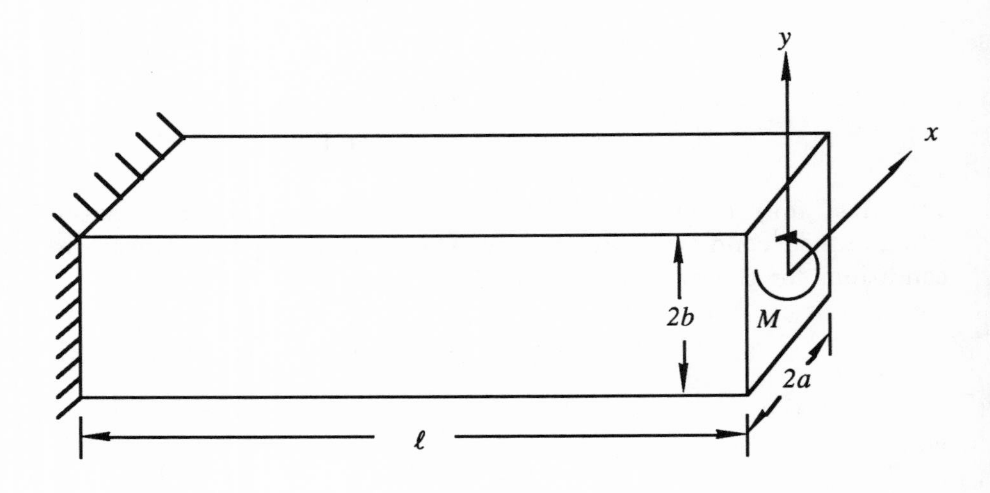

Figure 11.3. Torsional bar with rectangular cross section.

and

$$\tau_{zx}n_x + \tau_{zy}n_y = 0 \qquad \text{at } x = \pm a,\ y = \pm b. \tag{11.33}$$

Conditions (11.31) and (11.32) correspond to the first equation in (11.29), and the boundary conditions (11.33) correspond to the second equation in (11.29).

We choose a stress function $\psi(x,y)$ and define

$$\tau_{zx} = \frac{\partial \psi}{\partial y}, \quad \tau_{zy} = -\frac{\partial \psi}{\partial x} \tag{11.34}$$

so that the constraint condition (11.31) is satisfied. The boundary conditions (11.33) are

$$\frac{\partial \psi}{\partial y}n_x - \frac{\partial \psi}{\partial x}n_y = 0, \tag{11.35}$$

where $n_x = \partial y/\partial s$ and $n_x = -\partial x/\partial s$ along the boundary Γ cross section of the bar (see Fig. 10.3), or

$$\frac{\partial \psi}{\partial s} = 0, \qquad \psi = \text{constant on } \Gamma. \tag{11.36}$$

The constant in (11.36) can be taken as zero when Γ is a singly connected boundary. When the cross section has another boundary, $\psi = 1$ on that boundary. The condition (11.32) is written as

$$\begin{aligned} M &= -\int_{-b}^{b}\int_{-a}^{a}\left(\frac{\partial \psi}{\partial x}x + \frac{\partial \psi}{\partial y}y\right)dx\,dy \\ &= 2\int_{-b}^{b}\int_{-a}^{a}\psi\,dx\,dy. \end{aligned} \tag{11.37}$$

Then the variational principle is used to maximize

$$I^*[\psi] = -\frac{\ell}{2\mu}\int_{-b}^{b}\int_{-a}^{a}\left[\left(\frac{\partial \psi}{\partial x}\right)^2 + \left(\frac{\partial \psi}{\partial y}\right)^2\right]dx\,dy \tag{11.38}$$

with the boundary conditions

$$\psi = 0 \qquad \text{at } x = \pm a \text{ and } y = \pm b \tag{11.39}$$

and the constraint condition

$$M = 2\int_{-b}^{b}\int_{-a}^{a} \psi\, dx\, dy. \tag{11.40}$$

Choose

$$\psi = \sum_{m,n=1,3,5,\ldots} a_{mn} \cos\frac{m\pi x}{2a} \cos\frac{n\pi y}{2b} \tag{11.41}$$

so that conditions (11.39) are satisfied. Then

$$\begin{aligned} I^* &= -\frac{\ell}{2\mu}\sum_{m,n}\left(a_{mn}^2 \frac{m^2\pi^2}{4a^2} ab + a_{mn}^2 \frac{n^2\pi^2}{4b^2} ab \right), \\ M &= 2\sum_{mn} a_{mn} \frac{16ab}{mn\pi^2}(-1)^{(m+n)/2-1} = J. \end{aligned} \tag{11.42}$$

The values of a_{mn} are determined from

$$\begin{aligned} &\frac{\partial}{\partial a_{mn}}\left[I^* + \lambda(J - M)\right] = 0, \\ &\frac{\partial}{\partial \lambda}\left[I^* + \lambda(J - M)\right] = J - M = 0, \end{aligned} \tag{11.43}$$

which are

$$\begin{aligned} -\frac{\ell}{\mu}\left(\frac{m^2\pi^2}{4a^2} + \frac{n^2\pi^2}{4b^2}\right) ab a_{mn} + \lambda \frac{32ab}{mn\pi^2}(-1)^{(m+n)/2-1} = 0, \\ M = \sum_{m,n} a_{mn} \frac{32ab}{mn\pi^2}(-1)^{(m+n)/2-1} \end{aligned} \tag{11.44}$$

Then we have

$$\begin{aligned} a_{mn} &= \frac{\lambda\left[32(-1)^{(m+n)/2-1}/(mn\pi^2)\right]}{\ell/\mu\left[m^2\pi^2/(4a^2) + n^2\pi^2/(4b^2)\right]}, \\ M &= \lambda \sum_{m,n} \frac{32^2 ab/(m^2n^2\pi^4)}{\ell/\mu\left[m^2\pi^2/(4a^2) + n^2\pi^2/(4b^2)\right]}, \\ \lambda/(\ell/\mu) &= \frac{M}{\left\{\sum_{m,n}\left[32^2 ab/(m^2n^2\pi^4)\right]/\left[m^2\pi^2/(4a^2) + n^2\pi^2/(4b^2)\right]\right\}}. \end{aligned} \tag{11.45}$$

The problem of torsion in Fig. 11.3 is modified to the problem of finding M for a given twisting angle θ. The angle of torsion at the end of the bar is $\theta\ell$. In this case, Γ_2 in (11.27) compresses both ends of the bar. The boundary conditions are $\bar{u}_i = 0$ at $z = 0$ and $\bar{u}_1 = -\ell\theta y$, $\bar{u}_2 = \ell\theta x$, at $z = \ell$. Then, (11.27) becomes

$$I^*[\tau_{zx}, \tau_{zy}] = -\frac{\ell}{2\mu}\int_{-b}^{b}\int_{-a}^{a}\left(\tau_{zx}^2 + \tau_{zy}^2\right)dx\,dy + \ell\theta\int_{-b}^{b}\int_{-a}^{a}\left(\tau_{zy}x - \tau_{zx}y\right)dx\,dy. \quad (11.46)$$

The second term in (11.46) is the work done by the moment at the end of the bar and corresponds to the second term in (11.27).

The constraint conditions for the maximum problem of (11.46) are (11.31) and (11.33). Since M, or equivalently F, in (11.29) is not given, the domain Γ_1, where the force of the moment is given, vanishes in this problem. The condition (11.31) is satisfied by choosing the stress function ψ defined by (11.34). Then (11.46) becomes

$$I^*[\psi] = -\frac{\ell}{2\mu}\int_{-b}^{b}\int_{-a}^{a}\left[\left(\frac{\partial\psi}{\partial x}\right)^2 + \left(\frac{\partial\psi}{\partial y}\right)^2\right]dx\,dy + 2\ell\theta\int_{-b}^{b}\int_{-a}^{a}\psi\,dx\,dy. \quad (11.47)$$

The stress function ψ should satisfy (11.36), which corresponds to (11.33). The maximum of $I^*[\psi]$ is sought with the constraint condition (11.36). As an approximation for ψ, we take

$$\psi = \mu\theta\left(a^2 - x^2\right)\left(b^2 - y^2\right)\left(A_1 + A_2x^2 + A_3y^2 + \cdots\right) \quad (11.48)$$

that satisfies the constraint condition $\psi = 0$ on the boundary $x = \pm a$ and $y = \pm b$.

As a first approximation, we take

$$\psi = \mu\theta\left(a^2 - x^2\right)\left(b^2 - y^2\right)A_1. \quad (11.49)$$

When it is substituted into (11.47), we have

$$I^*(A_1) = \left[-\tfrac{64}{45}a^3b^3\left(a^2 + b^2\right)A_1^2 + \tfrac{32}{9}a^3b^3A_1\right]\mu\ell\theta^2. \quad (11.50)$$

The condition $\partial I^*/\partial A_1 = 0$ leads to

$$A_1 = \frac{5}{4(a^2+b^2)}. \tag{11.51}$$

The corresponding moment M is calculated from (11.37), giving

$$M = \frac{40}{9}\mu\theta\frac{a^3b^3}{a^2+b^2}. \tag{11.52}$$

When $a=b$, $M=0.1389(2a)^4\mu\theta$. The exact solution is $M=0.1406(2a)^4\mu\theta$.

Example 11.3
Consider Example 8.1 (see Fig. 8.1). The complementary energy is

$$I^*[\sigma_{ij}] = -\tfrac{1}{2}\iint_G (\sigma_x\varepsilon_x + \sigma_y\varepsilon_y + 2\sigma_{xy}\varepsilon_{xy})\,dx\,dy, \tag{11.53}$$

where

$$\begin{aligned}\varepsilon_x &= \frac{1}{E}(\sigma_x - \nu\sigma_y),\\ \varepsilon_y &= \frac{1}{E}(\sigma_y - \nu\sigma_x),\\ \varepsilon_{xy} &= \frac{1+\nu}{E}\sigma_{xy},\end{aligned} \tag{11.54}$$

ν is Poisson's ratio, and E is Young's modulus. I^* is rewritten as

$$I^*[\sigma_{ij}] = -\frac{1}{2E}\iint_G [\sigma_x^2 + \sigma_y^2 - 2\nu\sigma_x\sigma_y + 2(1+\nu)\sigma_{xy}^2]\,dx\,dy, \tag{11.55}$$

where σ_x, σ_y, and σ_{xy} must satisfy

$$\begin{aligned}\frac{\partial\sigma_x}{\partial x} + \frac{\partial\sigma_{xy}}{\partial y} &= 0,\\ \frac{\partial\sigma_{yx}}{\partial x} + \frac{\partial\sigma_y}{\partial y} &= 0,\end{aligned} \tag{11.56}$$

and the boundary conditions

$$\begin{aligned} \sigma_x &= s\left(1-\frac{y^2}{b^2}\right) && \text{at } x=\pm a, \\ \sigma_{xy} &= 0 && \text{at } x=\pm a,\ y=\pm b, \\ \sigma_y &= 0 && \text{at } y=\pm b. \end{aligned} \tag{11.57}$$

Choose ψ as

$$\sigma_x = \frac{\partial^2\psi}{\partial y^2}, \quad \sigma_y = \frac{\partial^2\psi}{\partial x^2}, \quad \sigma_{xy} = -\frac{\partial^2\psi}{\partial x\,\partial y}, \tag{11.58}$$

and

$$\psi = \frac{1}{2}sy^2\left(1-\frac{y^2}{6b^2}\right) + \left(x^2-a^2\right)^2\left(y^2-b^2\right)^2\left(a_1+a_2x^2+a_3y^2+\cdots\right). \tag{11.59}$$

Then conditions (11.56) and (11.57) are satisfied. The functional I^* is now

$$I^*[\psi] = -\frac{1}{2E}\iint_G\left[\left(\frac{\partial^2\psi}{\partial x^2}\right)^2 + \left(\frac{\partial^2\psi}{\partial y^2}\right)^2 - 2\nu\frac{\partial^2\psi}{\partial x^2}\frac{\partial^2\psi}{\partial y^2} + 2(1+\nu)\left(\frac{\partial^2\psi}{\partial x\partial y}\right)^2\right]dx\,dy. \tag{11.60}$$

When (11.59) is substituted into (11.60) and $\partial I^*/\partial a_1 = 0, \partial I^*/\partial a_2 = 0, \ldots$ are solved, the same result as in Example 8.1 is obtained.

Returning to expression (11.25), we will introduce the **Hellinger-Reissner principle** (E. Hellinger, "Der allgemeine Ansatz der Mechanik der Kontinua," *Encyclopdie der Mathematischen Wissenshaften*, vol. 4, part 4, pp.602–694, 1914; E. Reissner, "On a Variational Theorem in Elasticity," *J. Math. Phys.* 29, pp.90–95, 1950). Since $\sigma_{ij} = \lambda_{ij}$ at the stationary point, we can write (11.25) as

$$I\left[u_i,\varepsilon_{ij},\sigma_{ij},\lambda_i\right] = \tfrac{1}{2}\iiint_G C_{ijkl}\varepsilon_{kl}\varepsilon_{ij}\,dv - \iiint_G X_iu_i\,dv - \iint_{\Gamma_1} F_iu_i\,ds - \iiint_G \sigma_{ij}\left[\varepsilon_{ij} - \tfrac{1}{2}\left(u_{i,j}+u_{j,i}\right)\right]dv - \iint_{\Gamma_2}\lambda_i\left(u_i-\bar{u}_i\right)ds, \tag{11.61}$$

where F_i is a given force on Γ_1, $\bar{u}_i$ is a given displacement on Γ_2, and X_i is a given body force in G. The independent arguments of the functional are u_i, ε_{ij}, σ_{ij}, and λ_i. The physical meaning of λ_i is the reaction force on Γ_2.

The Hellinger and Reissner functional is obtained from (11.61) by eliminating ε_{ij} by Hookes law,

$$\varepsilon_{ij} = C_{ijkl}^{-1}\sigma_{kl}. \tag{11.62}$$

Then

$$I[u_i,\sigma_{ij},\lambda_i] = \frac{1}{2}\iiint_G C_{ijkl}^{-1}\sigma_{ij}\sigma_{kl}\,dv + \iiint_G \sigma_{ij}u_{i,j}\,dv - \iiint_G X_i u_i\,dv - \iint_{\Gamma_1} F_i u_i\,ds - \iint_{\Gamma_2} \lambda_i(u_i - \bar{u}_i)\,ds. \tag{11.63}$$

The Hellinger-Reissner principle considers that the real displacement and stress caused by the given F_i, X_i, $\bar{u}_i$ are obtained by the stationary functions of (11.63). The independent arguments are u_i, σ_{ij}, and λ_i, and no constraint conditions are required.

When the condition

$$\lambda_i = \sigma_{ij}n_j \tag{11.64}$$

in (11.26) is imposed, on integration by parts, (11.63) is written as

$$I[u_i,\sigma_{ij}] = -\frac{1}{2}\iiint_G C_{ijkl}^{-1}\sigma_{ij}\sigma_{kl}\,dv + \iiint_G (\sigma_{ij,j} + X_i)u_i\,dv + \iint_{\Gamma_1}(\sigma_{ij}n_j - F_i)u_i\,ds + \iint_{\Gamma_2}\sigma_{ij}n_j\bar{u}_i\,ds. \tag{11.65}$$

This is the free variational problem with independent arguments u_i and σ_{ij}.

If constraint conditions

$$\sigma_{ij,j} + X_i = 0 \quad \text{in } G,$$
$$\sigma_{ij}n_j = F_i \quad \text{on } \Gamma_1$$

are imposed, the functional becomes identical to (11.27).

Problems

11.1 A thin plate consisting of two materials whose Young's modulus and Poisson's ratio are E_1, ν_1 and E_2, ν_2 is subjected to an elongation u_1^0 at the end and fixed to the wall (see Fig. P11.1). Evaluate an upper bound and a lower bound of the elastic strain energy of the plate.

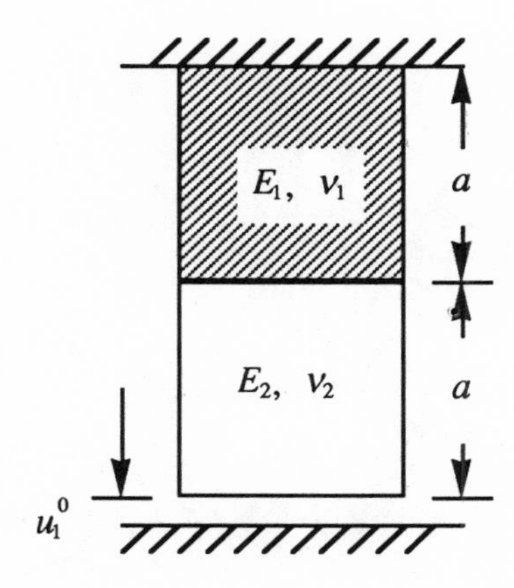

Figure P11.1. Thin plate consisting of two different materials.

11.2 A circular ring is composed of three layers. The core has thickness h, Young's modulus E_c and Poisson's ratio ν_c, and the two face layers have thickness t, Young's modulus E, Poisson's ratio ν. The ring is subjected to uniform normal forces, p_1 and p_2, and uniform shear forces, q_1 and q_2, per unit area on the inner and outer surfaces (see Fig. P11.2). When $h \ll 1$ and $Et/E_c h$ is relatively large, derive the following equations of equilibrium for the resultant moment M, stress resultant N, and resultant shear force Q;

$$\frac{dN}{d\theta} + Q + qa = 0,$$

$$\frac{dQ}{d\theta} - N + pa = 0,$$

$$\frac{dM}{d\theta} - aQ + ma = 0,$$

where

$$p = (1-\beta)p_1 - (1+\beta)p_2,$$
$$q = (1-\beta)q_1 + (1+\beta)q_2,$$
$$m = \beta a\left[(1+\beta)q_2 - (1-\beta)q_1\right],$$
$$\beta = \frac{h+t}{2a}.$$

What are the pertinent boundary conditions?

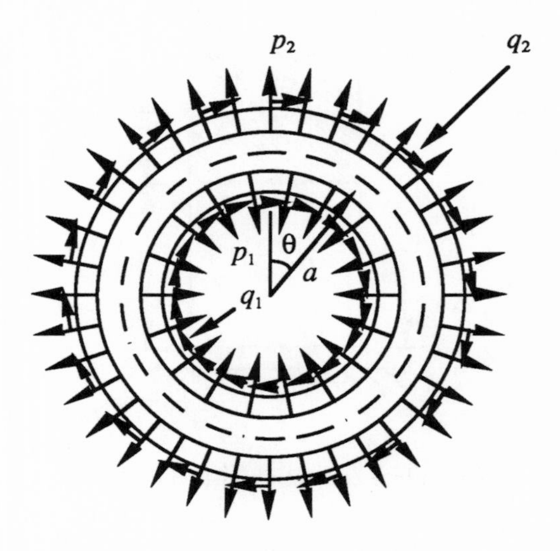

Figure P11.2. Cylinder ring composed of three layers.

11.3 An infinitely long circular cylinder with thickness t and internal radius a is subjected to internal pressure p, as shown in Fig. P11.3. By the use of the variational method, derive the equilibrium equation and the boundary conditions for stress components σ_r and σ_θ expressed in the polar coordinate system. The strain components ε_r and ε_θ are expressed with the displacement u in the radius direction, $\varepsilon_r = \partial u/\partial r$ and $\varepsilon_\theta = u/r$. Assume $\varepsilon_z = 0$.

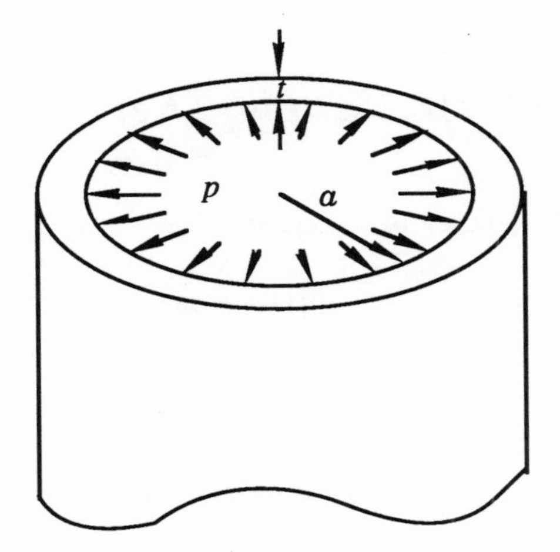

Figure P11.3. Cylinder with internal radius a and thickness t subjected to internal pressure p.

11.4 The total potential energy for a semi-infinite cylindrical shell subjected to a constant shear Q_0. and bending moment M_0, at $x = 0$, is

$$v = 2\pi a \int_0^{\infty} \left\{ \frac{Gh}{1-\nu} \left[\left(\frac{du}{dx} \right)^2 + 2\nu \left(\frac{du}{dx} \right) \frac{w}{a} + \frac{w^2}{a^2} \right] + \frac{Gh^3}{12(1-\nu)} \left(\frac{d^2u}{dx^2} \right)^2 \right\} dx$$

$$- 2\pi a Q_0 (w)_{x=0} + M_0 \left(\frac{dw}{dx} \right)_{x=0},$$

where $w(x)$ is the radial deflection of the shell, $u(x)$ the axial deflection, G the shear modulus, and ν Poisson's ratio; see Fig. P11.4. Find the equations governing u and w and the associated boundary conditions.

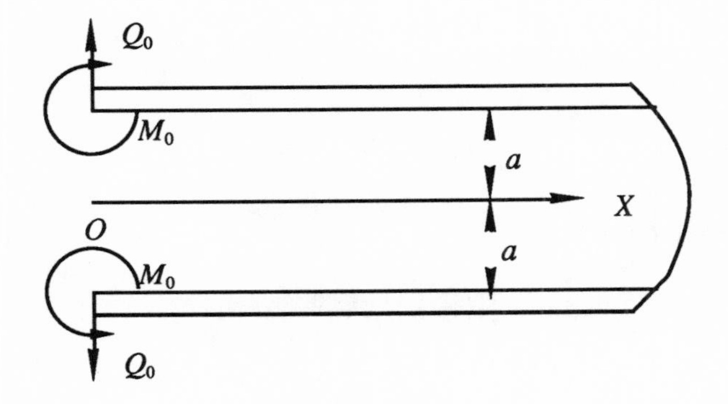

Figure P11.4. Cylindrical shell subjected to shear force Q_0 and moment M_0.

12

Castigliano's Theorem

The elastic strain energy of an elastic body is expressed as

$$W = \tfrac{1}{2} \iiint_G \sigma_{ij} \varepsilon_{ij} \, dv, \tag{12.1}$$

where

$$\sigma_{ij} = C_{ijkl} \varepsilon_{kl} \tag{12.2}$$

or

$$\varepsilon_{ij} = C^{-1}_{ijkl} \sigma_{kl}. \tag{12.3}$$

The tensor C^{-1}_{ijkl} is called the **compliance tensor**. Due to the property of symmetry of C_{ijkl} or C^{-1}_{ijkl}, the variation of (12.1) is written in two ways:

$$\delta W = \iiint_G \sigma_{ij} \delta \varepsilon_{ij} \, dv \tag{12.4}$$

or

$$\delta W = \iiint_G \varepsilon_{ij} \delta \sigma_{ij} \, dv. \tag{12.5}$$

When $\delta \varepsilon_{ij}$ is compatible, i.e.,

$$\delta \varepsilon_{ij} = \tfrac{1}{2} \left(\delta u_{i,j} + \delta u_{j,i} \right), \tag{12.6}$$

(12.4) is rewritten as

$$\begin{aligned} \delta W &= \iiint_G \sigma_{ij} \delta u_{i,j} \, dv \\ &= \iint_\Gamma \sigma_{ij} n_j \delta u_i \, ds - \iiint_G \sigma_{ij,j} \delta u_i \, dv, \end{aligned} \tag{12.7}$$

where Γ is the boundary of G. If

$$\begin{aligned} \sigma_{ij,j} &= 0 \qquad \text{in } G, \\ \sigma_{ij,j} n_j &= F_i \qquad \text{on } \Gamma, \end{aligned} \tag{12.8}$$

we have **Castigliano's theorem,**

$$\delta W[u_i] = \iint_\Gamma F_i \delta u_i \, ds. \tag{12.9}$$

It is emphasized that when (12.9) is used the strain energy must be expressed in terms of displacements.

Example 12.1.
Find the displacement at the point where P is applied in the truss as shown in Fig. 12.1, where cross sections of three members are S.

Let Δ be the displacement at point A. The strain in member AB or AD is $\Delta\cos\alpha/\ell$ (see Fig. 12.2), and that in member CA is $\Delta/(\ell\cos\alpha)$.

W is evaluated as

$$W[\Delta] = \frac{E}{2}\left(\frac{\Delta^2\cos^2\alpha}{\ell^2} 2\ell S + \frac{\Delta^2}{\ell^2\cos^2\alpha} S\ell\cos\alpha\right) \tag{12.10}$$

and

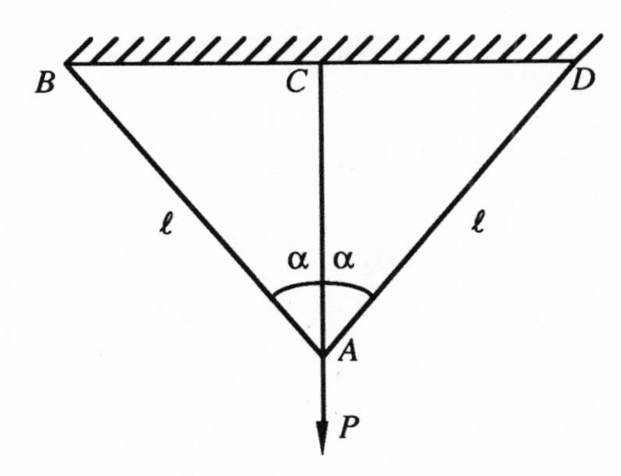

Figure 12.1. Truss composed of three bars AB, AC, and AD.

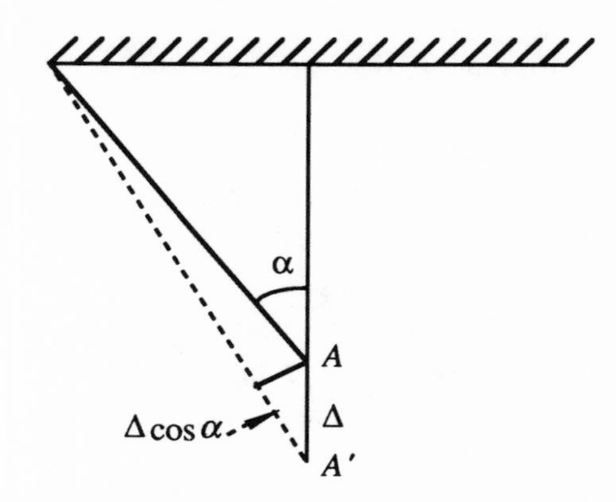

Figure 12.2. Geometry showing displacement Δ and strain; compatibility condition.

$$\delta W = E\left(\frac{\cos^2\alpha}{\ell}2S + \frac{S}{\ell\cos\alpha}\right)\Delta\delta\Delta.$$

The right-hand side of (12.9) is $P\delta\Delta$. Equation (12.9) becomes $\delta W = P\delta\Delta$. Then we have

$$P = E\left(\frac{\cos^2\alpha}{\ell}2S + \frac{S}{\ell\cos\alpha}\right)\Delta. \tag{12.11}$$

The uniformity of stresses in each member guarantees the conditions in (12.8).

Another Castigliano's theorem is obtained from (12.5). When ε_{ij} is compatible, i.e.,

$$\varepsilon_{ij} = \tfrac{1}{2}\left(u_{i,j} + u_{j,i}\right), \tag{12.12}$$

(12.15) is rewritten as

$$\begin{aligned}\delta W\left[\sigma_{ij}\right] &= \iiint_G \delta\sigma_{ij}u_{i,j}\,dv \\ &= \iint_\Gamma \delta\sigma_{ij}n_j u_i\,ds - \iiint_G \delta\sigma_{ij,j}u_i dv\end{aligned} \tag{12.13}$$

If

$$\begin{aligned}\sigma_{ij,j} &= 0 \quad \text{in } G, \\ \sigma_{ij}n_j &= F_i \quad \text{on } \Gamma.\end{aligned} \tag{12.14}$$

we have the complementary **Castigliano's theorem,**

$$\delta W[\sigma_{ij}] = \iint_{\Gamma} \delta F_i u_i \, ds \tag{12.15}$$

where W must be expressed in terms of stresses (or equivalent moments).

Example 12.2.

Find the deflection at the point where P is applied in the beam, as shown in Fig. 12.3.

The strain energy of the beam is

$$W = \frac{1}{2EI}\int_0^{\ell} M^2 \, dx, \tag{12.16}$$

where

$$M(x) = P(\ell - x). \tag{12.17}$$

Then

$$W = \frac{1}{2EI}\int_0^{\ell} P^2(\ell - x)^2 \, dx = \frac{P^2\ell^3}{6EI}. \tag{12.18}$$

Applying (12.15), we have

$$\delta W[M] = \frac{P\ell^3}{3EI}\delta P = (u)_{x=\ell}\delta P. \tag{12.19}$$

Thus, we have

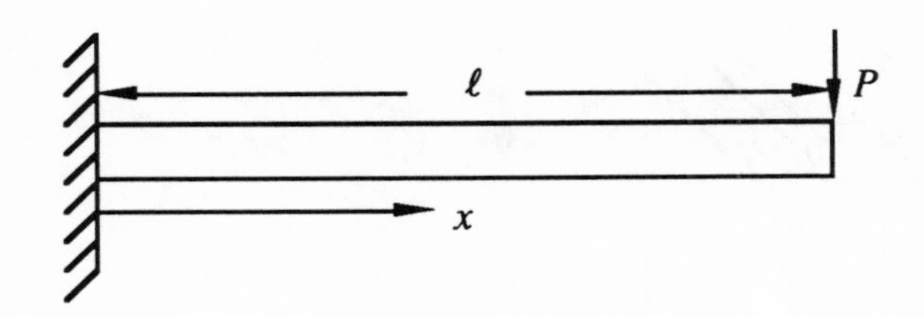

Figure 12.3. Clamped beam under force P.

$$(u)_{x=\ell} = \frac{P\ell^3}{3EI}. \tag{12.20}$$

In derivation, $M = (\ell - x)P$ corresponds to the conditions in (12.14).

Equation (12.15) can be applied even if δF_i is a fictitious or *dummy* force. By doing this, we can evaluate u_i in the direction of δF_i at the point where δF_i is defined.

Example 12.3.

A load P is applied at B to the structure shown in Fig. 12.4. Determine the horizontal and vertical deflection of point B.

We apply a dummy horizontal load Q at B. From the free-body diagram, we obtain the forces in members BC and BD,

$$\begin{aligned} F_{BC} &= 0.6P + 0.8Q, \\ F_{BD} &= -0.8P + 0.6Q. \end{aligned} \tag{12.21}$$

The elastic strain energy is

$$W = \frac{F_{BC}^2 a}{2AE} + \frac{F_{BD}^2 b}{2AE}. \tag{12.22}$$

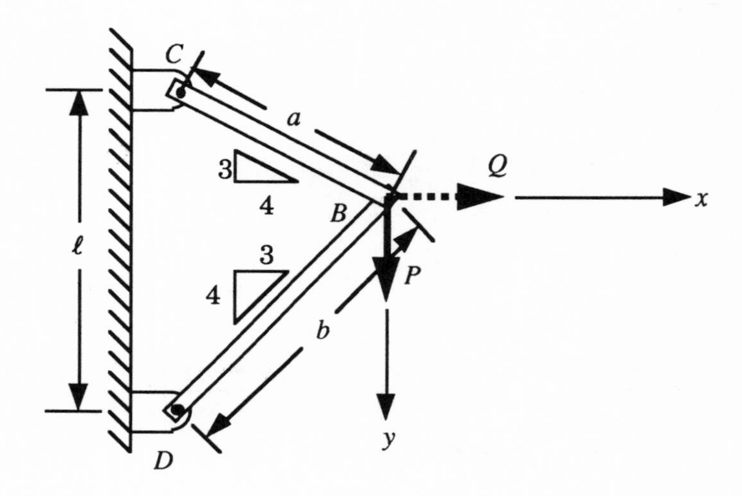

Figure 12.4. Truss subjected to force P and fictitious force Q

where A is the cross-section of the two rods. From Castigliano's theorem, (12.15), we have

$$\left(\frac{\partial W}{\partial Q}\right)_{Q\to 0} = u_1,$$
$$\left(\frac{\partial W}{\partial P}\right)_{Q\to 0} = u_2, \tag{12.23}$$

and, therefore,

$$u_1 = -0.096\frac{P\ell}{AE},$$
$$u_2 = 0.728\frac{P\ell}{AE}. \tag{12.24}$$

13

Plasticity

The Deformation Theory

The stress, strain, and displacement in a given body V must satisfy the following conditions for a given surface load F_i on Γ_1. The prescribed displacements on Γ_2 are

$$\sigma_{ij,j} = 0 \qquad \text{in } V, \tag{13.1.1}$$

$$\varepsilon_{ij} = \tfrac{1}{2}\left(u_{i,j} + u_{j,i}\right) \quad \text{in } V, \tag{13.1.2}$$

$$\sigma_{ij} = \sigma_{ij}\left(\varepsilon_{kl}\right) \qquad \text{in } V, \tag{13.1.3}$$

or conversely

$$\varepsilon_{ij} = \varepsilon_{ij}\left(\sigma_{kl}\right) \qquad \text{in } V, \tag{13.1.3.1}$$

$$\sigma_{ij} n_j = F_i \qquad \text{on } \Gamma_1, \tag{13.1.4}$$

$$u_i = \overline{u}_i \qquad \text{on } \Gamma_2. \tag{13.1.5}$$

The constitutive equation (13.1.3) (the relation between the stress and strain) is different from elasticity. The stress-strain relation is only valid for the loading case. When such a relation is proposed, the theory is called the **deformation theory**. This theory is good for work-hardening materials. For perfectly plastic materials, an alternative equation is used, because there is no one-to-one correspondence between the stress and the strain. When σ_{ij} satisfies (13.1.1) and (13.1.4) and satisfies (13.1.5), we have

$$\iiint_V \sigma_{ij} \delta u_{i,j} \, dv - \iint_{\Gamma_1} F_i \delta u_i \, ds = 0 \tag{13.2}$$

and

$$\iiint_V \tfrac{1}{2}\left(u_{i,j} + u_{j,i}\right) \delta\sigma_{ij} \, dv - \iint_{\Gamma_2} \overline{u}_i \delta\sigma_{ij} n_j \, ds = 0, \tag{13.3}$$

or

$$\iiint_V \sigma_{ij}\delta\varepsilon_{ij}\,dv - \iint_{\Gamma_1} F_i\delta u_i\,ds = 0 \tag{13.4}$$

and

$$\iiint_V \varepsilon_{ij}\delta\sigma_{ij}\,dv - \iint_{\Gamma_2} \overline{u}_i\delta\sigma_{ij}n_j\,ds = 0 \tag{13.5}$$

where ε_{ij} is compatible, satisfying (13.1.2).

In the theory of elasticity, we have had Castigliano's theorem, (12.9) or (12.15). It is hoped that a similar theorem may be derived for plasticity from (13.4) or (13.5). In order to do that, strain energy expressions similar to (12.1), (12.4), and (12.5) must be introduced.

If (13.1.3) are analytic functions that assure the existence of the state function $A(u_i)$, defined by

$$\delta A = \sigma_{ij}\delta\varepsilon_{ij}, \tag{13.6}$$

principle (13.4) leads to the principle of stationary potential energy

$$\delta I[u_i] = 0 \tag{13.7}$$

where

$$I[u_i] = \iiint_V A(u_i)\,dv - \iint_{\Gamma_1} F_i u_i\,ds. \tag{13.8}$$

If (13.1.3.1) are analytic functions that assure the existence of the state function $B(\sigma_{ij})$, defined by

$$\delta B = \varepsilon_{ij}\delta\sigma_{ij}, \tag{13.9}$$

the principle (13.5) leads to the principle of stationary complementary energy

$$\delta I^*[\sigma_{ij}] = 0 \tag{13.10}$$

where

$$I^*\left[\sigma_{ij}\right]=\iiint\limits_V B\left(\sigma_{ij}\right)dv-\iint\limits_{\Gamma_2}\bar{u}_i\sigma_{ij}n_j\,ds. \tag{13.11}$$

Let us consider the construction of the state functions $A(u_i)$ and $B(\sigma_{ij})$. According to the deformation theory called the **secant modulus theory**, we have

$$\sigma'_{ij}=\mu\varepsilon'_{ij} \tag{13.12}$$

where $\mu>0$, σ'_{ij}, and ε'_{ij} are the **stress and strain deviators** (**reduced stress and strain**), defined as

$$\begin{aligned}\sigma'_{ij}&=\sigma_{ij}-\delta_{ij}\sigma\\ \varepsilon'_{ij}&=\varepsilon_{ij}-\delta_{ij}\varepsilon\\ \sigma&=\tfrac{1}{3}\sigma_{kk}\\ \varepsilon&=\tfrac{1}{3}\varepsilon_{kk}.\end{aligned} \tag{13.13}$$

If the assumption (13.12) holds, we have

$$S=\mu\gamma \tag{13.14}$$

where

$$S=\sqrt{\sigma'_{ij}\sigma'_{ij}}\qquad \gamma=\sqrt{\varepsilon'_{ij}\varepsilon'_{ij}} \tag{13.15}$$

and consequently,

$$\begin{aligned}Sds&=\sigma'_{ij}d\sigma'_{ij}\\ \gamma d\gamma&=\varepsilon'_{ij}d\varepsilon'_{ij}\\ \mu&=\frac{S}{\gamma}.\end{aligned} \tag{13.16}$$

It is assumed that S is a single-valued continuous function of γ,

$$S=S(\gamma) \tag{13.17}$$

or conversely

$$\gamma=\gamma(S). \tag{13.17.1}$$

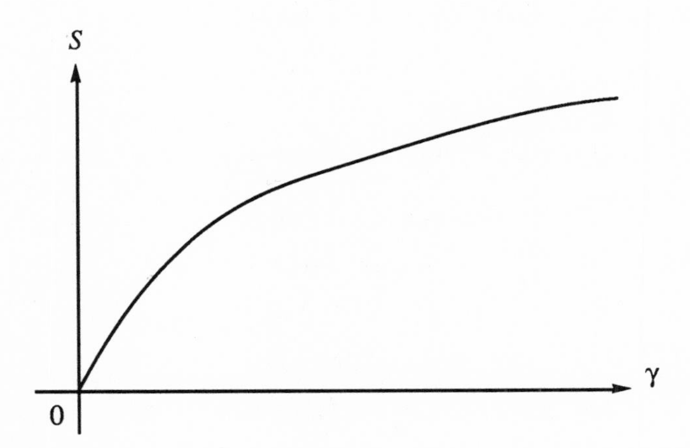

Figure 13.1. Experimental diagram of the shear stress and shear strain.

The above relation can be obtained from experiments of simple tension tests as in Fig. 13.1, where the tensile stress $\bar{\sigma}$ is equivalent to

$$\bar{\sigma} = \sqrt{\tfrac{3}{2}\left(\sigma'_{ij}\sigma'_{ij}\right)} = \sqrt{\tfrac{3}{2}}S \tag{13.18}$$

and the tensile strain $\bar{\varepsilon}$ is equivalent to

$$\bar{\varepsilon} = \sqrt{\tfrac{3}{2}\left(\varepsilon'_{ij}\varepsilon'_{ij}\right)} = \sqrt{\tfrac{3}{2}}\gamma . \tag{13.19}$$

In other words, the tension test, $\bar{\sigma} = \bar{\sigma}(\bar{\varepsilon})$, gives the relations (13.17). $\bar{\sigma}$ is variously known as the **generalized stress, effective stress**, or **equivalent stress**. $\bar{\varepsilon}$ is the **generalized strain, effective strain,** or **equivalent strain.** The basic philosophy of the equivalent stress and strain comes from the fact that the expressions of (13.18) and (13.19) are reduced to the tensile stress and strain as special cases, where $\sigma_{11} = \bar{\sigma}$, $\sigma_{22} = \sigma_{33} = 0$, and $\sigma_{11} = \bar{\sigma}$, $\varepsilon_{11} = \bar{\varepsilon}$, $\varepsilon_{22} = \varepsilon_{33} = -\bar{\varepsilon}/2$. The state functions A and B are obtained as

$$A(u_i) = \frac{9}{2}K\varepsilon^2 + \int_0^{\gamma} S(\gamma)\,d\gamma, \tag{13.20}$$

$$B(\sigma_{ij}) = \frac{1}{2K}\sigma^2 + \int_0^{S} \gamma(S)\,dS \tag{13.21}$$

where K is the bulk modulus of the material,

$$\sigma = K\varepsilon_{mm}. \tag{13.22}$$

The reason for the above construction is obvious when we consider

$$\begin{aligned}
\delta A &= 9K\varepsilon\delta\varepsilon + S(\gamma)\delta\gamma \\
&= K\varepsilon_{mm}\delta\varepsilon_{ii} + \mu\gamma\delta\gamma \\
&= K\varepsilon_{mm}\delta_{ij}\delta\varepsilon_{ij} + \mu\varepsilon'_{ij}\delta\varepsilon'_{ij} \\
&= \left(K\varepsilon_{mm}\delta_{ij} + \sigma'_{ij}\right)\delta\varepsilon_{ij} \\
&= \left(\sigma\delta_{ij} + \sigma'_{ij}\right)\delta\varepsilon_{ij} \\
&= \sigma_{ij}\delta\varepsilon_{ij},
\end{aligned} \tag{13.23}$$

and

$$\begin{aligned}
\delta B &= \frac{1}{K}\sigma\delta\sigma + \gamma\delta S \\
&= \frac{1}{K}\sigma\delta\sigma + \gamma\frac{\sigma'_{ij}\delta\sigma'_{ij}}{S} \\
&= \frac{1}{3}\varepsilon_{mm}\delta_{ij}\delta\sigma_{ij} + \frac{1}{\mu}\sigma'_{ij}\delta\sigma'_{ij} \\
&= \left(\varepsilon\delta_{ij} + \varepsilon'_{ij}\right)\delta\sigma_{ij} \\
&= \varepsilon_{ij}\delta\sigma_{ij},
\end{aligned} \tag{13.24}$$

which lead to (13.6) and (13.9).

Kachanov's Principles

Principle 1: The exact solution of the problem of the deformation theory of plasticity renders the functional $I[u_i]$, defined in (13.8), a minimum with respect to admissible displacement variations.

It is proved as follows:

$$I[u_i + \delta u_i] = \iiint_V A(u_i + \delta u_i)dv - \iint_{\Gamma_1} F_i(u_i + \delta u_i)ds,$$

$$\begin{aligned} A(u_i + \delta u_i) &= \frac{9}{2}K(\varepsilon + \delta\varepsilon)^2 + \int_0^{\gamma+\Delta\gamma} S(\gamma)d\gamma \\ &= \frac{9}{2}K\left[\varepsilon^2 + 2\varepsilon\delta\varepsilon + (\delta\varepsilon)^2\right] + \int_0^{\gamma} S(\gamma)d\gamma + \int_{\gamma}^{\gamma+\Delta\gamma} S(\gamma)d\gamma \\ &\approx A(u_i) + 9K\varepsilon\delta\varepsilon + \frac{9}{2}K(\delta\varepsilon)^2 + S(\gamma)\Delta\gamma, \end{aligned} \tag{13.25}$$

$$\begin{aligned} \gamma + \Delta\gamma &= \sqrt{(\varepsilon'_{ij} + \delta\varepsilon_{ij})(\varepsilon'_{ij} + \delta\varepsilon'_{ij})} \\ &\approx \gamma + \frac{\varepsilon'_{ij} + \delta\varepsilon'_{ij}}{\gamma} + \frac{\delta\varepsilon'_{ij}\delta\varepsilon'_{ij}}{2\gamma} \\ &= \gamma + \delta\gamma + \frac{\delta\varepsilon'_{ij}\delta\varepsilon'_{ij}}{2\gamma}. \end{aligned}$$

Thus, we have

$$\begin{aligned} I[u_i + \delta u_i] - I[u_i] &= \iiint_V \left[9K\varepsilon\delta\varepsilon + S(\gamma)\delta\gamma\right]dv \\ &\quad - \iint_{\Gamma_1} F_i\delta u_i\, ds + \iiint_V \left[\frac{9}{2}K(\delta\varepsilon)^2 + S(\gamma)\frac{\delta\varepsilon'_{ij}\delta\varepsilon'_{ij}}{2\gamma}\right]dv \\ &= \iiint_V \left[\frac{9}{2}K(\delta\varepsilon)^2 + S(\gamma)\frac{\delta\varepsilon'_{ij}\delta\varepsilon'_{ij}}{2\gamma}\right]dv > 0 \end{aligned} \tag{13.26}$$

Principle 2: The exact solution of the problem of the deformation theory of plasticity renders the functional $I^*[\sigma_{ij}]$, defined in (13.11), a minimum with respect to admissible stress variations.

The proof can be done similarly to (13.25) and (13.26).

Perfectly Plastic Materials

The S-γ curve obtained from simple tension tests becomes Fig. 13.2 when the material is perfectly plastic. The problem of plasticity must satisfy all conditions in (13.1) except (13.1.3) or (13.1.3.1). Instead, we must impose the following conditions.

The strain is the sum of elastic e_{ij} and plastic ε^P_{ij},

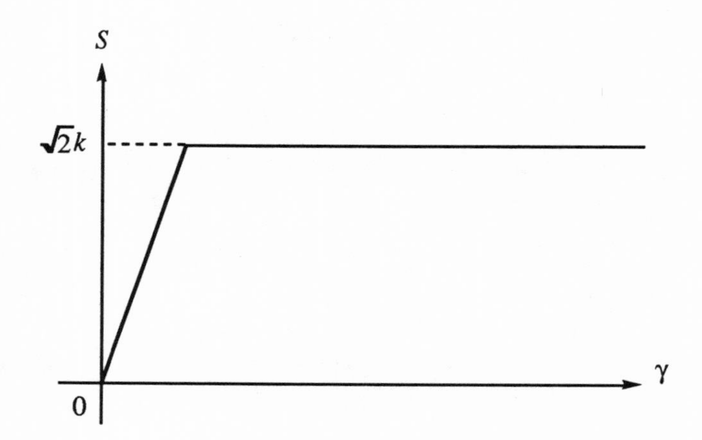

Figure 13.2. Stress-strain relation for perfectly plastic materials.

$$\varepsilon_{ij} = e_{ij} + \varepsilon_{ij}^{P} \tag{13.27}$$

and

$$e_{ij} = \frac{1}{3K}\sigma\delta_{ij} + \frac{1}{2G}\sigma'_{ij} \tag{13.28}$$

where G is the shear modulus. The plastic strain follows

$$\varepsilon_{ij}^{P} = \lambda\sigma'_{ij} \tag{13.29}$$

where $\lambda > 0$. When $\varepsilon_{ij}^{P} \neq 0$,

$$\sigma'_{ij}\sigma'_{ij} = 2k^2 \tag{13.30}$$

holds, and when $\varepsilon_{ij}^{P} = 0$,

$$\sigma'_{ij}\sigma'_{ij} < 2k^2. \tag{13.31}$$

The above conditions, (13.30) and (13.31), are called the **yielding conditions.** For the perfectly plastic materials, the following **Haar-Kármán's principle** holds. Among arbitrary sets of admissible stress components that satisfy the equations of equilibrium, the mechanical boundary conditions on Γ_1, and the yielding conditions, the exact solutions renders

$$I^*[\sigma_{ij}] = \iiint_V \left(\frac{1}{2K}\sigma^2 + \frac{1}{4G}\sigma'_{ij}\sigma'_{ij} \right) dv - \iint_{\Gamma_2} \sigma_{ij} n_j \bar{u}_i \, ds \tag{13.32}$$

a minimum.

The proof is as follows:

$$\begin{aligned}
\delta I^* &= \iiint_V \left(\frac{1}{K}\sigma\delta\sigma + \frac{1}{2G}\sigma'_{ij}\delta\sigma'_{ij} \right) dv - \iint_{\Gamma_2} \delta\sigma_{ij} n_j \bar{u}_i \, ds \\
&= \iiint_V e_{ij}\delta\sigma_{ij} \, dv - \iint_{\Gamma_2} \delta\sigma_{ij} n_j \bar{u}_i \, ds \\
&= \iiint_V \left(u_{i,j}\delta\sigma_{ij} - \varepsilon^P_{ij}\delta\sigma_{ij} \right) dv - \iint_{\Gamma_2} \delta\sigma_{ij} n_j \bar{u}_i \, ds \\
&= -\iiint_V \varepsilon^P_{ij}\delta\sigma_{ij} \, dv \\
&= -\iiint_V \lambda\sigma'_{ij}\delta\sigma_{ij} \, dv
\end{aligned} \tag{13.33}$$

The exact solution satisfies $\sigma'_{ij}\sigma'_{ij} = 2k^2$ in the plastic region, while the admissible solution has been chosen so that $\sigma'^*_{ij}\sigma'^*_{ij} \le 2k^2$. **Schwartz's inequality** proves that

$$\sigma'_{ij}\sigma'^*_{ij} \le \sqrt{\sigma'_{ij}\sigma'_{ij}}\sqrt{\sigma'_{ij}\sigma'_{ij}} \le 2k^2 \tag{13.34.1}$$

and, therefore,

$$\begin{aligned}
\sigma'_{ij}\delta\sigma_{ij} &= \sigma'_{ij}\left(\sigma^*_{ij} - \sigma_{ij}\right) \\
&= \sigma'_{ij}\sigma'^*_{ij} - \sigma'_{ij}\sigma'_{ij} \le 0.
\end{aligned} \tag{13.34.2}$$

Consequently, we conclude that

$$\delta I^* \ge 0. \tag{13.35}$$

When the S-γ diagram is simplified as in Fig. 13.3, the material is called the **Hencky material.** The conditions for the solution of the problem are the same as for the perfectly plastic materials except the elastic strain e_{ij} in (13.27) is ignored. The principle of stationary potential energy (13.7) holds, where

Figure 13.3. Stress-strain relation for Hencky materials.

$$
\begin{aligned}
A(u_i) &= \int_0^{\gamma} S(\gamma)\,d\gamma \\
&= \sqrt{2}k\sqrt{\varepsilon'_{ij}\varepsilon'_{ij}} \\
&= \sqrt{2}k\sqrt{\varepsilon_{ij}\varepsilon_{ij}}
\end{aligned}
\tag{13.36}
$$

under the **incompressible condition.** This principle becomes as follows. Among admissible solutions that satisfy the condition of compatibility, the geometrical boundary conditions on Γ_2, and the incompressibility condition, the actual solution renders

$$
I[u_i] = \sqrt{2}k\iiint_V \sqrt{\varepsilon_{ij}\varepsilon_{ij}}\,dv - \iint_{\Gamma_1} F_i u_i\,ds \tag{13.37}
$$

a minimum.

The principle of minimum complementary energy (13.10) holds where $B = 0$. Among admissible solutions that satisfy the equation of equilibrium, the yield condition, and the mechanical boundary condition on Γ_1, the actual solution renders

$$
I^*[\sigma_{ij}] = -\iint_{\Gamma_2} \bar{u}_i \sigma_{ij} n_j\,ds \tag{13.38}
$$

a minimum. This principle is equivalent to **Sadowsky's principle** of maximum plastic work.

The Flow Theory

At a given instance of time t, a body V is assumed to be in a state of static equilibrium. The stress σ_{ij}, the strain rate $\dot{\varepsilon}_{ij}$, and the velocity v_i satisfy the conditions,

$$\sigma_{ij,j} = 0, \qquad \dot{\varepsilon}_{ij} = \mu \sigma'_{ij}$$

when $\sigma'_{ij}\sigma'_{ij} = 2k^2$,

$$\dot{\varepsilon}_{ij} = 0$$

when $\sigma'_{ij}\sigma'_{ij} < 2k^2$

$$\begin{aligned} \dot{\varepsilon}_{ij} &= \tfrac{1}{2}\left(v_{i,j} + v_{j,i}\right), \\ \dot{\varepsilon}_{ii} &= 0, \\ \sigma_{ij} n_j &= F_i \qquad \text{on } \Gamma_1, \\ v_i &= \bar{v}_i \qquad \text{on } \Gamma_2. \end{aligned} \tag{13.39}$$

Markov's principle is as follows. Among admissible solutions that satisfy the condition of compatibility and incompressibility as well as the geometrical boundary conditions on Γ_2, the actual solution renders

$$I[u_i] = \sqrt{2}k \iiint_V \sqrt{\dot{\varepsilon}_{ij}\dot{\varepsilon}_{ij}}\, dv - \iint_{\Gamma_1} F_i v_i \, ds \tag{13.40}$$

a minimum.

The proof is as follows. Let the stress, strain rate, and velocity of the exact solution be denoted by σ_{ij}, $\dot{\varepsilon}_{ij}$, and v_i, and the strain rate and velocity of an admissible solution by $\dot{\varepsilon}^*_{ij}$ and v^*_i. Then

$$\begin{aligned}\sigma_{ij}\dot{\varepsilon}_{ij}^{*} &\le \sqrt{\sigma'_{ij}\sigma'_{ij}}\sqrt{\dot{\varepsilon}_{ij}^{*}\dot{\varepsilon}_{ij}^{*}} \le \sqrt{2}k\sqrt{\dot{\varepsilon}_{ij}^{*}\dot{\varepsilon}_{ij}^{*}} \\ \sigma_{ij}\dot{\varepsilon}_{ij} &= \mu\sigma'_{ij}\sigma'_{ij} = \frac{\sqrt{\dot{\varepsilon}_{ij}\dot{\varepsilon}_{ij}}}{\sqrt{\sigma'_{ij}\sigma'_{ij}}}\left(\sigma'_{ij}\sigma'_{ij}\right) \\ &= \sqrt{2}k\sqrt{\dot{\varepsilon}_{ij}\dot{\varepsilon}_{ij}}\end{aligned} \tag{13.41}$$

and, therefore,

$$\sqrt{2}k\sqrt{\dot{\varepsilon}_{ij}^{*}\dot{\varepsilon}_{ij}^{*}} - \sqrt{\dot{\varepsilon}_{ij}\dot{\varepsilon}_{ij}} \ge \sigma_{ij}\left(\dot{\varepsilon}_{ij}^{*} - \dot{\varepsilon}_{ij}\right). \tag{13.42}$$

Consequently, we have

$$\sqrt{2}k\iiint_V \sqrt{\dot{\varepsilon}_{ij}^{*}\dot{\varepsilon}_{ij}^{*}}\,dv - \iint_{\Gamma_1} F_i v_i^{*}\,ds \ge \sqrt{2}k\iiint_V \sqrt{\dot{\varepsilon}_{ij}\dot{\varepsilon}_{ij}}\,dv - \iint_{\Gamma_1} F_i v_i\,ds. \tag{13.43}$$

The second principle may be stated as follows. Among admissible solutions that satisfy the equations of equilibrium, the yielding condition, and the mechanical boundary conditions on Γ_1, the actual solution renders

$$I^{*}\left[\sigma_{ij}\right] = -\iint_{\Gamma_2} \sigma_{ij} n_j \bar{v}_i\,ds \tag{13.44}$$

a minimum.

The proof is as follows. Let the stress, strain rate, and velocity of the actual solution be denoted by σ_{ij}, $\dot{\varepsilon}_{ij}$, and v_i, and the stress of an admissible solution by σ_{ij}^{*}. Then, we have

$$\begin{aligned}&\sigma'_{ij}\sigma'_{ij} = 2k^2, \qquad \sigma_{ij}'^{*}\sigma_{ij}'^{*} = 2k^2 \\ &\sigma'_{ij}\sigma_{ij}'^{*} \le \sqrt{\sigma'_{ij}\sigma'_{ij}}\sqrt{\sigma_{ij}'^{*}\sigma_{ij}'^{*}} = 2k^2\end{aligned} \tag{13.45}$$

and, therefore,

$$\left(\sigma_{ij}'^{*} - \sigma'_{ij}\right)\sigma'_{ij} \le 0 \tag{13.46}$$

or

$$\left(\sigma^*_{ij} - \sigma_{ij}\right)\dot{\varepsilon}_{ij} \le 0. \tag{13.47}$$

Integrating (13.47), we have

$$\iint\limits_{\Gamma_2} \sigma^*_{ij} n_j \bar{v}_i \, ds \le \iint\limits_{\Gamma_2} \sigma_{ij} n_j \bar{v}_i \, ds. \tag{13.48}$$

Limit Analysis

Consider a perfectly plastic continuum or structure that is in equilibrium under a given force F_i on Γ_1, and constraint $\bar{u}_i = 0$ on Γ_2. The safety factor m of this system is determined from the condition that the system collapses under the load mF_i. At the state of the impending plastic collapse, the following conditions are satisfied:

$$\begin{aligned}
\sigma_{ij,j} &= 0, & & \\
\sigma'_{ij}\sigma'_{ij} &= 2k^2, & & \\
\dot{\varepsilon}_{ij} &= \mu\sigma'_{ij}, & &\text{where } \sigma'_{ij}\sigma'_{ij} = 2k^2, \\
\dot{\varepsilon}_{ij} &= 0 & &\text{where } \sigma'_{ij}\sigma'_{ij} < 2k^2, \\
\dot{\varepsilon}_{ij} &= \tfrac{1}{2}\left(v_{i,j} + v_{j,i}\right), & & \\
\dot{\varepsilon}_{ii} &= 0, & & \\
\sigma_{ij} n_j &= mF_i & &\text{on } \Gamma_1, \\
v_i &= 0 & &\text{on } \Gamma_2.
\end{aligned} \tag{13.49}$$

A set of stress components σ^*_{ij} will be called **statically admissible** if it satisfies

$$\begin{aligned}
\sigma^*_{ij,j} &= 0, \\
\sigma'^*_{ij}\sigma'^*_{ij} &\le 2k^2, \\
\sigma^*_{ij} n_j &= m_s F_i \qquad \text{on } \Gamma_1
\end{aligned} \tag{13.50}$$

where m_s is a number called a statically **admissible multiplier**.

A set of velocity components v^*_i will be called **kinematically admissible** if it satisfies

$$\begin{aligned} &\dot{\varepsilon}^*_{ij} = \tfrac{1}{2}\left(v^*_{i,j} + v^*_{j,i}\right), \\ &\dot{\varepsilon}^*_{ii} = 0, \\ &v^*_i = 0 \qquad\qquad \text{on } \Gamma_1, \\ &\int_{\Gamma_1} F_i v^*_i \, ds > 0. \end{aligned} \tag{13.51}$$

A number defined by

$$m_k = \sqrt{2}k \frac{\iiint_V \sqrt{\dot{\varepsilon}^*_{ij}\dot{\varepsilon}^*_{ij}} \, dv}{\int_{\Gamma_1} F_i v^*_i \, ds} \tag{13.52}$$

is called a **kinematically admissible multiplier.**

The **limit analysis theorem** is

$$m_s \leq m \leq m_k. \tag{13.53}$$

The proof is as follows. According to Schwartz's inequality,

$$\sigma'_{ij}\dot{\varepsilon}^*_{ij} \leq \sqrt{\sigma'_{ij}\sigma'_{ij}} \sqrt{\dot{\varepsilon}^*_{ij}\dot{\varepsilon}^*_{ij}} \tag{13.54}$$

or

$$\sigma'_{ij}\dot{\varepsilon}^*_{ij} \leq \sqrt{2}k\sqrt{\dot{\varepsilon}^*_{ij}\dot{\varepsilon}^*_{ij}}. \tag{13.55}$$

Integrating, we have

$$m\iint_{\Gamma_1} F_i v^*_i \, ds \leq \sqrt{2}k \iiint_V \sqrt{\dot{\varepsilon}^*_{ij}\dot{\varepsilon}^*_{ij}} \, dv \tag{13.56}$$

and, therefore, $m \leq m_k$. We have also

$$\left(\sigma^*_{ij} - \sigma_{ij}\right)\dot{\varepsilon}_{ij} \leq 0 \tag{13.57}$$

because it means

$$\left(\sigma^*_{ij} - \sigma_{ij}\right)\mu\sigma'_{ij} \leq 0 \tag{13.58}$$

or

$$\sigma_{ij}'^{*}\sigma_{ij}' - \sigma_{ij}'\sigma_{ij}' \leq 0 \tag{13.59}$$

where

$$\sigma_{ij}'^{*}\sigma_{ij}' \leq \sqrt{\sigma_{ij}'^{*}\sigma_{ij}'^{*}}\sqrt{\sigma_{ij}'\sigma_{ij}'} \leq 2k^2, \tag{13.60}$$

$$\sigma_{ij}'\sigma_{ij}' \leq 2k^2. \tag{13.61}$$

The subtraction (13.60)–(13.61) leads to (13.59), which is equivalent to (13.57). Integrating (13.57), we have $m_s - m \leq 0$.

14

Eigenvalue Problems

First, several eigenvalue problems are illustrated by examples before applications of the variational methods are discussed.

Example 14.1

A uniform string with length ℓ and density ρ is under tension p. What are the frequencies of the string when it vibrates?

The equation of motion is

$$p\frac{\partial^2 u}{\partial x^2} = \rho\frac{\partial^2 u}{\partial t^2}, \tag{14.1}$$

where u is the deflection of the string. Put

$$u = v(x)g(t). \tag{14.2}$$

Then

$$pv''g = \rho v\ddot{g} \tag{14.3}$$

or

$$\frac{pv''}{\rho v} = \frac{\ddot{g}}{g}. \tag{14.4}$$

The left-hand side of (14.4) is a function of x, and the right-hand side is a function of t. They are equal only when they are constant. The constant is denoted by $-\lambda$.

Then we have

$$\begin{aligned} pv'' + \lambda\rho v &= 0, \\ \ddot{g} + \lambda g &= 0. \end{aligned} \tag{14.5}$$

The second equation (14.5) has the solution

$$g = a\cos\sqrt{\lambda}t + b\sin\sqrt{\lambda}t \tag{14.6}$$

where $\sqrt{\lambda}$ is the frequency of vibration, and a and b are constants. The first equation of (14.5) has the solution

$$v = c_1 \sin\sqrt{\frac{\lambda\rho}{p}}x + c_2 \cos\sqrt{\frac{\lambda\rho}{p}}x. \tag{14.7}$$

The boundary conditions of the string are

$$u = 0 \text{ at } x = 0 \text{ and } \ell. \tag{14.8}$$

Therefore, $c_2 = 0$ and

$$\sqrt{\frac{\lambda\rho}{p}}\ell = n\pi, \quad (n = 1,2,3,\ldots). \tag{14.9}$$

Therefore, the required frequencies are $\sqrt{\lambda} = (n\pi/\ell)\sqrt{p/\rho}$ ($n = 1,2,3,\ldots$). λ is called the **eigenvalue,** and $v = c_1 \sin\sqrt{\lambda\rho/p}\,x$ is called the **eigenfunction.** $\lambda_n = (n\pi/\ell)^2(p/\rho)$ is the ***n*th eigenvalue** and $v_n = c_1 \sin\sqrt{\lambda_n\rho/p}\,x$ the ***n*th eigenfunction.** $\sqrt{\lambda_n}$ is the ***n*th eigenfrequency.**

By definition, we have

$$\begin{aligned} pv_n'' + \lambda_n\rho v_n &= 0, \\ pv_m'' + \lambda_m\rho v_m &= 0. \end{aligned} \tag{14.10}$$

Multiplying v_m by the first equation and v_n by the second equation, we have

$$\int_0^\ell (pv_n''v_m - pv_m''v_n)\,dx = (\lambda_m - \lambda_n)\int_0^\ell \rho v_n v_m\,dx. \tag{14.11}$$

On the other hand,

$$\int_0^\ell (pv_n''v_m - pv_m''v_n)\,dx = [pv_n'v_m]_0^\ell - [pv_m'v_n]_0^\ell = 0. \tag{14.12}$$

The differential equation is called the **self-adjoint,** when (14.12) is satisfied.

From (14.11) and (14.12) we see that

$$\int_0^\ell \rho v_n v_m \, dx = 0 \tag{14.13}$$

when $\lambda_n \neq \lambda_m$. Namely, the eigenfunctions are **orthogonal.**

Example 14.2

Find the eigenvalues and eigenfunctions for the differential equation

$$\frac{\partial^2 u}{\partial x^2} + \frac{\partial^2 u}{\partial y^2} + \lambda u = 0, \quad 0 \le r \le 1, \tag{14.14}$$

with the boundary condition

$$u = 0 \quad \text{on } r = 1, \tag{14.15}$$

where $r = \sqrt{x^2 + y^2}$.

The polar coordinate expression for (14.14) is

$$u_{rr} + \frac{1}{r} u_r + \frac{1}{r^2} u_{\theta\theta} + \lambda u = 0. \tag{14.16}$$

Put

$$u = v(r)(a \cos n\theta + b \sin n\theta). \tag{14.17}$$

Then (14.16) is written as

$$r^2 v'' + r v' + \left(r^2 \lambda - n^2\right) v = 0. \tag{14.18}$$

The above expression becomes, after transformation of variable, $\rho = r\sqrt{\lambda}$,

$$\frac{d^2 v}{d\rho^2} + \frac{1}{\rho} \frac{dv}{d\rho} + \left(1 - \frac{n^2}{\rho^2}\right) v = 0, \tag{14.19}$$

which has the Bessel functions as solutions,

$$v = J_n\left(r\sqrt{\lambda}\right). \tag{14.20}$$

The boundary conditions (14.15) is satisfied when

$$J_n\left(r\sqrt{\lambda}\right) = 0. \tag{14.21}$$

The roots of (14.21) are the eigenvalues of the problem. Denoting the eigenvalues by λ_{nm}, the corresponding eigenfunctions are written as

$$u_n = \sum_m J_n\left(r\sqrt{\lambda_{nm}}\right)\left(a_{nm}\cos n\theta + b_{nm}\sin n\theta\right). \tag{14.22}$$

Example 14.3
Consider a dynamical system with n degrees of freedom where kinetic and potential energy are expressed by

$$\begin{aligned} T &= a_{ij}\dot{q}_i\dot{q}_j, \\ U &= b_{ij}q_iq_j, \end{aligned} \tag{14.23}$$

where $a_{ij} = a_{ji}$ and $b_{ij} = b_{ji}$. Find the eigenvalues or eigenfrequencies associated with this system.

Consider a linear transformation

$$q_i(t) = A_{ij}n_j(t) \tag{14.24}$$

that transforms

$$\begin{aligned} a_{ij}A_{ik}A_{jl} &= \delta_{kl}, \\ b_{ij}A_{ik}A_{jl} &= \delta_{kl}\lambda_{(l)}, \end{aligned} \tag{14.25}$$

where the summation convention is not applied to $\delta_{kl}\lambda_{(l)}$. $A_{ij} = A_{ji}$ and δ_{ij} is the Kronecker delta. In order to find such a transformation and the values of λ_i, we first define

$$\begin{aligned} A(\mathbf{x},\mathbf{y}) &= a_{ij}x_iy_j, \\ B(\mathbf{x},\mathbf{y}) &= b_{ij}x_iy_j. \end{aligned} \tag{14.26}$$

Next, consider the problem to minimize

$$\frac{B(\mathbf{x},\mathbf{x})}{A(\mathbf{x},\mathbf{x})} \tag{14.27}$$

where $A(\mathbf{x},\mathbf{x})$ is assumed to be positive definite. The stationary point is denoted by $\mathbf{x}=\mathbf{x}^1$, or $x_1 = x_1^1$, $x_2 = x_2^1$, $x_3 = x_3^1,\ldots$. The minimum value is denoted by λ_1. We write

$$\begin{aligned} B(\mathbf{x}^1,\mathbf{x}^1) &= \lambda_1, \\ A(\mathbf{x}^1,\mathbf{x}^1) &= 1 \end{aligned} \tag{14.28}$$

without losing generality.

Next minimize

$$\frac{B(\mathbf{x},\mathbf{x})}{A(\mathbf{x},\mathbf{x})} \tag{14.29}$$

with the constant condition

$$A(\mathbf{x},\mathbf{x}^1) = 0. \tag{14.30}$$

The stationary point is denoted by $\mathbf{x}=\mathbf{x}^2$ or $x_1 = x_1^2$, $x_2 = x_2^2$, $x_3 = x_3^2,\ldots,$ and the minimum value is denoted by λ_2. Then

$$\begin{aligned} B(\mathbf{x}^2,\mathbf{x}^2) &= \lambda_2, \\ A(\mathbf{x}^2,\mathbf{x}^2) &= 1, \\ A(\mathbf{x}^2,\mathbf{x}^1) &= 0. \end{aligned} \tag{14.31}$$

Next minimize

$$\frac{B(\mathbf{x},\mathbf{x})}{A(\mathbf{x},\mathbf{x})} \tag{14.32}$$

with constraint conditions

$$
\begin{aligned}
A(\mathbf{x},\mathbf{x}^1) &= 0, \\
A(\mathbf{x},\mathbf{x}^2) &= 0.
\end{aligned}
\tag{14.33}
$$

The process continues until λ_n is defined. The seeking coefficients A_{ij} of the linear transformation (14.24) are found as

$$
A_{ij} = x_i^j. \tag{14.34}
$$

The reasoning is obvious for the following relations:

$$
\begin{aligned}
B(\mathbf{x}^1,\mathbf{x}^1) &= \lambda_1 \\
&= b_{ij}x_i^1x_j^1 = b_{ij}A_{i1}A_{j1} \\
A(\mathbf{x}^1,\mathbf{x}^1) &= 1 \\
&= a_{ij}x_i^1x_j^1 = a_{ij}A_{i1}A_{j1} \\
A(\mathbf{x}^2,\mathbf{x}^1) &= 0 \\
&= a_{ij}x_i^2x_j^1 = a_{ij}A_{i2}A_{j1} \\
B(\mathbf{x}^2,\mathbf{x}^2) &= \lambda_2 \\
&= b_{ij}x_i^2x_j^2 = b_{ij}A_{i2}A_{j2} \\
A(\mathbf{x}^2,\mathbf{x}^2) &= 1 \\
&= a_{ij}x_i^2x_j^2 = a_{ij}A_{i2}A_{j2}, \text{ etc.}
\end{aligned}
\tag{14.35}
$$

The relations in (14.35) are nothing more than those in (14.25).

By using the linear transformation (14.24), we write

$$
\begin{aligned}
T &= \delta_{ij}\dot{n}_i\dot{n}_j, \\
U &= \lambda n_i^2.
\end{aligned}
\tag{14.36}
$$

The Euler equations associated with (14.36) are

$$
n_i + \lambda_{(i)}n_i = 0, \tag{14.37}
$$

where the summation convention is not applied to $\lambda_{(i)}n_i$ and $\lambda_{(1)}n_1 = \lambda_1 n_1$, $\lambda_{(2)}n_2 = \lambda_2 n_2, \ldots$. λ_i is the ith eigenvalue, and $\sqrt{\lambda_i}$ is the ith eigenfrequency. λ_i is larger than λ_{i-1} since the ith minimum of $B(\mathbf{x},\mathbf{x})/A(\mathbf{x},\mathbf{x})$ is obtained under the constraint conditions for λ_i, which has one more constraint condition than λ_{i-1},

$$\lambda_1 \leq \lambda_2 \leq \cdots \leq \lambda_n. \tag{14.38}$$

The eigenvalue` λ can also be interpreted as a minimum value of $B(\mathbf{x},\mathbf{x})$ under constraint condition $A(\mathbf{x},\mathbf{x})=1$. By the use of the Lagrange multiplier λ, the variation of $B(\mathbf{x},\mathbf{x})-\lambda A(\mathbf{x},\mathbf{x})$ is considered. Then

$$\delta B(\mathbf{x},\mathbf{x}) - \lambda \delta A(\mathbf{x},\mathbf{x}) = 0 \tag{14.39}$$

leads to

$$b_{ij}x_j - \lambda a_{ij}x_j = 0. \tag{14.40}$$

The nontrivial values of $x_1, x_2, \ldots$ exist only when the determinant is zero,

$$\begin{vmatrix} b_{11}-\lambda a_{11} & b_{12}-\lambda a_{12} & \cdots & b_{1n}-\lambda a_{1n} \\ b_{21}-\lambda a_{21} & b_{22}-\lambda a_{22} & \cdots & b_{2n}-\lambda a_{2n} \\ \vdots & \vdots & \ddots & \vdots \\ b_{n1}-\lambda a_{n1} & b_{n2}-\lambda a_{n2} & \cdots & b_{nn}-\lambda a_{nn} \end{vmatrix} = 0. \tag{14.41}$$

Multiplying x_i by (14.40), we have

$$\lambda = \frac{b_{ij}x_i x_j}{a_{kl}x_k x_l}. \tag{14.42}$$

Therefore, λ is a minimum value of $b_{ij}x_i x_j$ when $a_{ij}x_i x_j$ is taken as 1.

15

Variational Principles and Eigenvalues

We consider the eigenvalue problem associated with

$$\frac{\partial^2 u}{\partial x^2}+\frac{\partial^2 u}{\partial y^2}+\lambda u=0 \quad \text{in } G, \qquad u=0 \quad \text{on } \Gamma. \tag{15.1}$$

where Γ is the boundary of G. The boundary condition must be homogeneous, and $\partial u/\partial n=0$ can also be considered instead of $u=0$.

The minimum problem of the functional

$$I[u]=\iiint_G\left[\left(\frac{\partial u}{\partial x}\right)^2+\left(\frac{\partial u}{\partial y}\right)^2\right]dx\,dy \tag{15.2}$$

with the constraint conditions

$$H[u]=\iint_G u^2\,dx\,dy=1 \tag{15.3}$$

and with the boundary condition $u=0$ on Γ leads to the Euler equation in (15.1). λ is the Lagrange multiplier for (15.3). Namely, the variational method corresponding to (15.1) leads to the minimum problem of the functional

$$J[u]=\iint_G\left[\left(\frac{\partial u}{\partial x}\right)^2+\left(\frac{\partial u}{\partial y}\right)^2\right]dx\,dy-\lambda\iint_G u^2\,dx\,dy \tag{15.4}$$

with the boundary condition $u=0$ on Γ. Show that $J=0$ for the stationary function and $\lambda=I[u]/H[u]$.

Since $I[u]$ and $H[u]$ have homogeneous integrands of the same order, the variational principle used to minimize (15.2) with the constraint condition (15.3) is equivalent to minimizing the functional $I[u]/H[u]$. If $H[u]$ is not 1, there exists a constant c such that $H[cu]=1$ and $I[u]/H[u]=I[cu]/H[cu]=I[cu]$. The minimum value of $I[u]/H[c]$ is denoted by λ_1,

$$\frac{I[u]}{H[u]} \geq \lambda_1. \tag{15.5}$$

Denoting the first eigenfunction by u_1, we define the second eigenvalue and eigenfunction by the variational principle to minimize

$$\frac{I[u]}{H[u]} \tag{15.6}$$

with the constraint condition

$$H[u,u_1] = 0, \tag{15.7}$$

where

$$H[u,u_1] = \iint_G uu_1 \, dx \, dy. \tag{15.8}$$

The minimum value of (15.6) is λ_2, and the stationary function is u_2. Similarly, we can define the ith eigenvalue by the minimum problem of (15.6) with the constraint conditions

$$\begin{aligned} H[u,u_1] &= 0, \\ H[u,u_2] &= 0, \\ &\vdots \\ H[u,u_{i-1}] &= 0. \end{aligned} \tag{15.9}$$

The homogeneous boundary condition $u = 0$ on Γ is imposed for all of these variational problems.

The definition of λ_i mentioned above assumes that the eigenfunctions $u_1, u_2, \ldots, u_{i-1}$ are known. In practice this is not the case. Therefore, the following **maximum-minimum principle** is employed.

We find a minimum of

$$I[\psi] = \iint_G \left[\left(\frac{\partial \psi}{\partial x} \right)^2 + \left(\frac{\partial \psi}{\partial y} \right)^2 \right] dx \, dy \tag{15.10}$$

with the constraint conditions

$$H[\psi]=\iint_G \psi^2\,dx\,dy=1 \tag{15.11}$$

and

$$\begin{aligned}
H[\psi,v_1]&=\iint_G \psi v_1\,dx\,dy=0,\\
H[\psi,v_2]&=\iint_G \psi v_2\,dx\,dy=0,\\
&\vdots\\
H[\psi,v_{i-1}]&=\iint_G \psi v_{i-1}\,dx\,dy=0,
\end{aligned} \tag{15.12}$$

where $v_1, v_2,\ldots, v_{i-1}$ are given piecewise continuous functions. The minimum value $\overline{\lambda}_i$ may depend on the choice of functions $v_1, v_2,\ldots, v_{i-1}$. We will show that λ_i is the maximum of $\overline{\lambda}_i$ with respect to $v_1, v_2,\ldots, v_{i-1}$.

Thus,

$$\lambda_i=\max\overline{\lambda}_i. \tag{15.13}$$

Let us consider a linear combination

$$\psi=c_1u_1+c_2u_2+\cdots+c_iu_i, \tag{15.14}$$

where $u_1, u_2,\ldots, u_i$ are the eigenfunctions. When it is substituted into (15.10), we have

$$I[\psi]=\sum_{k=1}^{i}c_k^2\lambda_k. \tag{15.15}$$

When (15.14) is substituted into (15.11),

$$\sum_{k=1}^{i}c_k^2=1. \tag{15.16}$$

Since $\overline{\lambda}_i$ is the minimum of $I[\psi]$ and $\lambda_i \geq \lambda_1, \lambda_2,\ldots, \lambda_{i-1}$, we have

$$\overline{\lambda}_i\leq I[\psi]=c_1^2\lambda_1+c_2^2\lambda_2+\cdots+c_i^2\lambda_i\leq\left(\lambda_i\sum_{k=1}^{i}c_k^2\right)=\lambda_i, \tag{15.17}$$

which proves the relation (15.13).

16

Direct Methods for Eigenvalue Problems

In this chapter several eigenvalue problems are solved by direct methods. Eigenvalue problems are strongly related to dynamic problems. In many cases they are more difficult to analyze than static problems so that analytical approaches are not practical. Use of direct methods is in order. Ritz's method in Chapter 3 and Galerkin's method in Chapter 8 are equally applicable.

Ritz's Method

A common form of a Sturm-Liouville boundary-value problem is

$$
\begin{aligned}
\frac{d}{dx}\left[p(x)\frac{du}{dx}\right]+[\lambda\rho(x)-q(x)]u=0, \qquad & a\le x\le b\\
\alpha_1 u(a)+\alpha_2 u'(a)=0,&\\
\beta_1 u(b)+\beta_2 u'(b)=0,&\\
p(x)>0,&\\
\rho(x)>0,&\\
\alpha_1^2+\alpha_2^2\neq 0,&\\
\beta_1^2+\beta_2^2\neq 0&
\end{aligned}
\tag{16.1}
$$

where $p(x)$, $\rho(x)$, and $q(x)$ are continuous on the interval $a\le x\le b$. In order to apply Ritz's method to find the eigenvalues associated with this problem, the functional form must be derived. From the definition of the eigenvalue the differential equation in (16.1) is rewritten as

$$
\frac{d}{dx}\left[p(x)\frac{du}{dx}\right]+[\lambda\rho(x)-q(x)]u=-\omega u
$$

or

$$
\frac{d}{dx}\left[p(x)\frac{du}{dx}\right]+[(\lambda\rho(x)+\omega)-q(x)]u=0. \tag{16.2}
$$

The functional of (16.2) can be constructed as follows:

$$\int_a^b \left\{ \frac{d}{dx}\left[p(x)\frac{du}{dx} \right] + \{[\lambda \rho(x) + \omega] - q(x)\}u \right\} \delta u \, dx = 0$$
$$p(x)\frac{du}{dx}\delta u \Big|_a^b - \int_a^b \left\{ p(x)\frac{du}{dx}\delta u_x - \{\{\lambda \rho(x) + \omega\} - q(x)\}u\delta u \right\} dx = 0. \tag{16.3}$$

The integral term in the second line of (16.3) is the variation of

$$I[u] = \frac{1}{2}\int_a^b \left\{ p(x)\left(\frac{du}{dx}\right)^2 - \{\{\lambda \rho(x) + \omega\} - q(x)\}u^2 \right\} dx. \tag{16.4}$$

This is the functional to be used. To apply Ritz's method assume a trial function of the form

$$u(x) \cong \sum_{j=1}^{N} c_j \varphi_j(x). \tag{16.5}$$

Substitution of this trial function into (16.4) yields

$$I[u] \cong \frac{1}{2}\int_a^b \left\{ p(x)\left[\sum_{j=1}^{N} c_j \left(\frac{d\varphi_j}{dx} \right) \right]^2 - \{[\lambda \rho(x) + \omega] - q(x)\} \left(\sum_{j=1}^{N} c_j \varphi_j \right)^2 \right\} dx. \tag{16.6}$$

This equation can be written as

$$I[u] \cong T(c_1, \ldots, c_N, \omega, \lambda). \tag{16.7}$$

(16.7) is extremized with respect to each constant c_j; the result is

$$\frac{\partial T}{\partial c_j} = 0, \quad j = 1, \ldots, N. \tag{16.8}$$

This is a system of simultaneous equations. Because of the functional's quadratic nature, all the c_j's in T are either quadratic or bilinear; therefore, (16.8) can be rewritten as

$$\frac{\partial T}{\partial c_i} c_i = A_{ij} c_j = 0. \tag{16.9}$$

In order for the above equation to have a nontrivial solution, the determinant of the matrix in (16.9) is zero, that is,

$$\det\left|A_{ij}\right| = f(\omega, f) = 0. \tag{16.10}$$

The determinant in (16.10) is called the **frequency equation**. The process of finding the eigenvalues is reduced to searching the roots of the frequency equation. These roots correspond to the eigenvalues of the problem.

Notice that the number of eigenvalues found depends on the number of terms N taken. If two terms have been chosen, there are two roots in the frequency equation.

Example 16.1

Find the eigenvalues associated with the problem

$$\begin{aligned} &\frac{\partial^2 u}{\partial x^2} + u = 0, \qquad 0 \le x \le 1; \\ &u(0) = u(1) = 0. \end{aligned} \tag{16.11}$$

This is a special case of $p(x) = \rho(x) = \lambda = 1$ and $q(x) = 0$. By (16.4) the functional for this problem is

$$I[u] = \frac{1}{2}\int_0^1 \left[\left(\frac{du}{dx}\right)^2 - (1+\omega)u^2\right] dx. \tag{16.12}$$

Assume the trial function

$$u(x) \cong x(x-1)(a_1 + a_2 x). \tag{16.13}$$

When this trial function is substituted into (16.12) the resulting functional (16.7) is

$$T = \frac{9-\omega}{60}a_1^2 + \frac{9-\omega}{60}a_1 a_2 + \frac{13-\omega}{210}a_2^2. \tag{16.14}$$

When this equation is extremized with respect to each constant, (16.14) becomes

$$A_{ij}a_j = \begin{bmatrix} \dfrac{9-\omega}{30} & \dfrac{9-\omega}{60} \\ \dfrac{9-\omega}{60} & \dfrac{13-\omega}{105} \end{bmatrix} \begin{Bmatrix} a_1 \\ a_2 \end{Bmatrix} = 0. \tag{16.15}$$

Therefore, the frequency equation is

$$f(\omega) = \det\left|A_{ij}\right| = \omega^2 - 50\omega + 369 = 0. \tag{16.16}$$

The roots are

$$\begin{aligned} \omega &= 9; \\ \omega &= 41. \end{aligned} \tag{16.17}$$

The exact solutions are

$$\begin{aligned} \omega &= \pi^2 - 1 \cong 8.86974; \\ \omega &= 4\pi^2 - 1 \cong 38.47842. \end{aligned} \tag{16.18}$$

Table 16.1 shows the results of using up to four terms.

As can be seen from Table 16.1, the larger the eigenvalue, the worse the accuracy. Extra terms must be included in the trial function to compute smaller eigenvalues.

Using Mathematica

Table 16.1 can be easily recreated using Mathematica (see the Appendix for details).

Table 16.1. Eigenvalues by Ritz's method.

Trial function: $u(x) \cong x(x-1)\sum_{j=1}^{N} a_j x^{j-1}$

N	ω_1	ω_2	ω_3	ω_4
1	9.00000			
2	9.00000	41.00000		
3	8.86974	41.00000	101.13025	
4	8.86974	38.50155	101.13025	199.49845
Exact	8.86960	38.47842	87.82644	156.91367

Galerkin's Method
Example 16.2
Find the eigenvalues of the eigenvalue problem

$$\frac{d}{dx}[p(x)y'] - qy + \lambda y = 0, \quad (p > 0,\ q \geq 0) \tag{16.19}$$

with the boundary conditions

$$y(0) = y(\ell) = 0 \tag{16.20}$$

by the use of Galerkin's method.
Assume

$$y = \sum_{n=1}^{N} a_n \sin\frac{n\pi x}{\ell} \tag{16.21}$$

so that it satisfies (16.20). Substitute it into

$$\int_0^\ell \left[\frac{d}{dx}(py') - qy + \lambda y\right] \sin\frac{n\pi x}{\ell}\, dx = 0, \quad n = 1, 2, \ldots. \tag{16.22}$$

Let

$$\begin{aligned} \int_0^\ell \left[\left(p\frac{m\pi}{\ell}\cos\frac{m\pi x}{\ell}\right)' - q\sin\frac{m\pi x}{\ell}\right]\sin\frac{n\pi x}{\ell}\, dx &= \alpha_{mn}, \\ \int_0^\ell \sin\frac{m\pi x}{\ell}\sin\frac{n\pi x}{\ell}\, dx &= \frac{\delta_{mn}\ell}{2}. \end{aligned} \tag{16.23}$$

Then (16.22) becomes

$$\sum_{m=1}^{N} a_m\left(\alpha_{mn} + \frac{\lambda\delta_{mn}\ell}{2}\right) = 0, \quad n = 1, 2, 3, \ldots. \tag{16.24}$$

Nontrivial values of a_m exist only when

$$\begin{vmatrix} \alpha_{11}+\frac{\lambda\ell}{2} & \alpha_{12} & \cdots & \alpha_{1N} \\ \alpha_{21} & \alpha_{22}+\frac{\lambda\ell}{2} & \cdots & \alpha_{2N} \\ \vdots & \vdots & \ddots & \vdots \\ \alpha_{N1} & \alpha_{N2} & \cdots & \alpha_{NN}+\frac{\lambda\ell}{2} \end{vmatrix} = 0. \tag{16.25}$$

The roots of λ in (16.25) are approximate solutions for $\lambda_1, \lambda_2, \ldots, \lambda_N$.

Example 16.3

The differential equation

$$\nabla^2 v + \lambda v = 0 \tag{16.26}$$

is defined in a circle with radius a and with the boundary condition $v = 0$ at $r = a$. Find the eigenvalue λ by a direct method of the variational principle. Assume that the eigenvalues are symmetrical. Then (16.26) is written as

$$\frac{1}{r}\frac{d}{dr}\left(r\frac{dv}{dr}\right) + \lambda v = 0. \tag{16.27}$$

Choose the fundamental functions

$$w_1 = \cos\left(\frac{\pi r}{2a}\right), \quad w_2 = \cos\left(\frac{3\pi r}{2a}\right), \quad w_3 = \cos\left(\frac{5\pi r}{2a}\right), \quad \cdots \tag{16.28}$$

satisfying the boundary condition $v = 0$ at $r = a$.

The first eigenfunction is approximated by

$$v_1 = a_1 \cos\left(\frac{\pi r}{2a}\right). \tag{16.29}$$

According to Galerkin's method, we have

$$\int_0^a \left[\frac{1}{r}\frac{d}{dr}\left(r\frac{dv_1}{dr}\right) + \lambda v_1\right] w_1\, 2\pi r\, dr = 0, \tag{16.30}$$

or

$$\frac{\pi^2}{4}\left(\frac{1}{2}+\frac{2}{\pi^2}\right)-\lambda a^2\left(\frac{1}{2}-\frac{2}{\pi^2}\right)=0. \tag{16.31}$$

The first eigenvalue is approximated as

$$\lambda_1 = \frac{5.830}{a^2}. \tag{16.32}$$

The exact value of λ_1 is known to be $\lambda_1 = 5.779/a^2$.

The second approximation of the first eigenfunction is assumed to be

$$v_1 = a_1 \cos\left(\frac{\pi r}{2a}\right) + a_2 \cos\left(\frac{3\pi r}{2a}\right). \tag{16.33}$$

Then

$$\begin{aligned}
&\int_0^a \left[\frac{1}{r}\frac{d}{dr}\left(r\frac{dv_1}{dr}\right)+\lambda v_1\right] w_1\, 2\pi r\, dr = 0,\\
&\int_0^a \left[\frac{1}{r}\frac{d}{dr}\left(r\frac{dv_1}{dr}\right)+\lambda v_1\right] w_2\, 2\pi r\, dr = 0,
\end{aligned} \tag{16.34}$$

or numerically

$$\begin{aligned}
&(1.7337-0.29736a)a_1 + \left(0.20264\lambda a^2 - 1.5\right)a_2 = 0\\
&\left(0.20264a^2 - 1.5\right)a_1 + \left(11.603 - 0.47748\lambda a^2\right)a_2 = 0.
\end{aligned} \tag{16.35}$$

The zero determinant of the matrix of (16.35) leads to

$$0.10092\lambda^2 a^4 - 3.6702\lambda a^2 + 17.866 = 0. \tag{16.36}$$

The smallest root of (16.36) is

$$\lambda_1 = \frac{5.790}{a^2}. \tag{16.37}$$

The error is estimated as 0.19% when compared with the exact solution. When the value of λ_1 given by (16.19) is substituted into one of the equations in (16.35), we have

$$0.0144a_1 - 0.3263a_2 = 0 \tag{16.38}$$

or

$$a_1 = 0.3263C, \quad a_2 = 0.0114C \tag{16.39}$$

with the proportionality constant C. The approximated first eigenfunction, therefore, is obtained as

$$u_1 = C\left[0.3263\cos\left(\frac{\pi r}{2a}\right) + 0.0114\cos\left(\frac{3\pi r}{2a}\right)\right]. \tag{16.40}$$

Problems

16.1 Find the eigenfrequency of the torsional system shown in Fig. P16.1. The moments of inertia of the bar and the disk about the axis are H and H_1, respectively. The torsional rigidity of the bar is C.

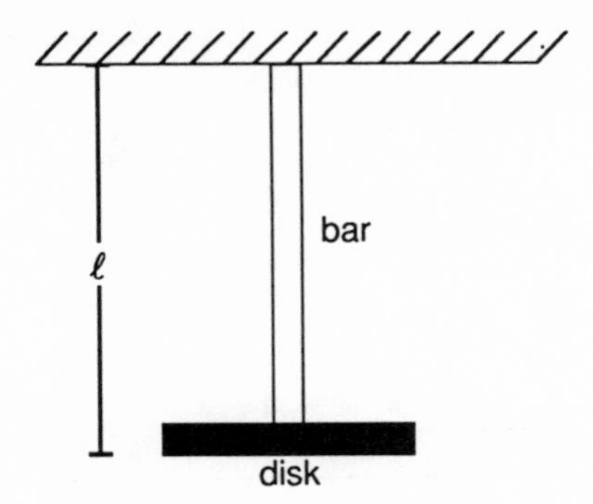

Figure P16.1. Torsion system.

16.2 Apply the Galerkin or Ritz's method to solve the eigenvalue problem

$$\Delta v + \lambda v = 0$$

define in a circular domain with radius a and with the boundary condition

$$\frac{\partial v}{\partial r} = 0 \quad \text{at } r = a.$$

Hint:

$$\Delta v + \lambda v = 0 \rightarrow \frac{1}{r}\frac{d}{dr}\left(r\frac{dv}{dr}\right) + \lambda v = 0,$$

$$v = a_1 \sin\left(\frac{\pi r}{2a}\right) + a_2 \sin\left(\frac{3\pi r}{2a}\right) + \cdots.$$

16.3 Two disks are attached to a torsional bar which is built in the two walls, as shown in Fig. P16.3. Find the lowest eigenfrequency of the system, where the torsional rigidity of the bar is C, the moment of inertia per unit length of the bar is I, and the two disks have moments of inertia I_1 I_2, respectively. For mathematical simplicity, assume that $\ell_2 = 2\ell_1$, $\ell_3 = 3\ell_1$, $I_1 = \ell_1 I$, and $I_2 = 2\ell_1 I$.

Hint:

$$T = \frac{1}{2}\int_0^{\ell_3} I\dot{\psi}^2(x,t)\,dt + \frac{1}{2}I_1\dot{\psi}^2(\ell_1,t) + \frac{1}{2}I_2\dot{\psi}^2(\ell_2,t),$$

$$U = \frac{1}{2}\int_0^{\ell_3} C(\psi')^2\,dx.$$

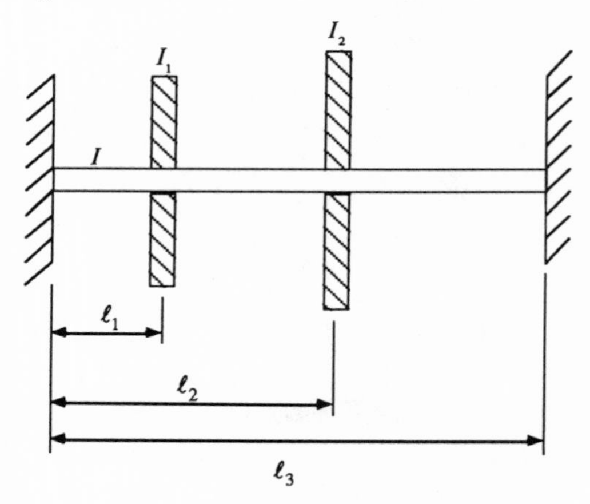

Figure P16.3. Two disks attached to a torsional bar.

16.4 Determine the first eigenvalue λ_1 for

$$\frac{d}{dx}\left(\sqrt{1+x}\,\frac{du}{dx}\right) + \lambda u = 0, \quad 0 \le x \le 1,$$

$$u(0) = u(1) = 0.$$

17

The Finite Element Method

Let us take solving the Poisson equation as an example

$$\frac{\partial^2 \varphi}{\partial x^2} + \frac{\partial^2 \varphi}{\partial y^2} + C = 0 \tag{17.1}$$

in a given domain G with the boundary condition $\varphi = \varphi_0$ on Γ. This is equivalent to finding a function that satisfies the boundary condition and minimizes the functional

$$I[\varphi] = \iint_G \frac{1}{2}\left[\left(\frac{\partial \varphi}{\partial x}\right)^2 + \left(\frac{\partial \varphi}{\partial y}\right)^2\right] dx\,dy - \iint_G C\varphi\,dx\,dy. \tag{17.2}$$

The domain is divided into finite elements with triangular shapes as shown in Fig. 17.1. The typical triangular element has modes i, j, and m numbered in a counterclockwise order. The coordinates of point i are denoted by x_i and y_i those of point j by x_j and y_j, and those of point m by x_m and y_m.

The trial function φ in (17.2) is expressed by a linear function of x and y in each triangle,

$$\varphi = \alpha_1 + \alpha_2 x + \alpha_3 y, \tag{17.3}$$

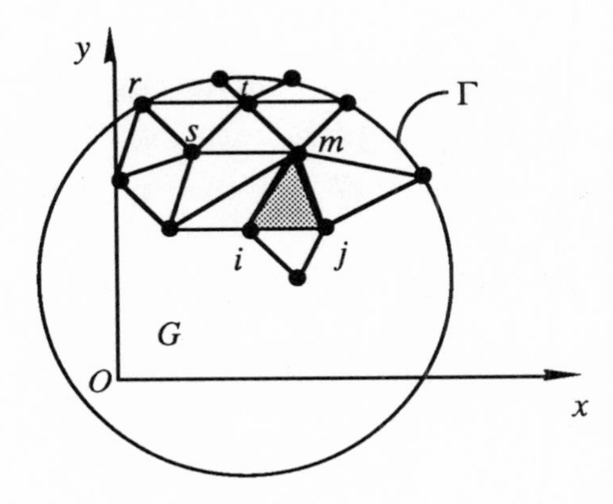

Figure 17.1. Triangular finite elements.

where parameters α_1, α_2, and α_3 are different in each triangle. The values at the nodal points are

$$\begin{aligned}\varphi_i &= \alpha_1 + \alpha_2 x_i + \alpha_3 y_i,\\ \varphi_j &= \alpha_1 + \alpha_2 x_j + \alpha_3 y_j,\\ \varphi_m &= \alpha_1 + \alpha_2 x_m + \alpha_3 y_m.\end{aligned} \tag{17.4}$$

We express α_1, α_2, α_3 in terms of φ_i, φ_j, φ_m, and the nodal point coordinates. Then

$$\begin{aligned}\alpha_1 &= \frac{a_i\varphi_i + a_j\varphi_j + a_m\varphi_m}{2\Delta},\\ \alpha_2 &= \frac{b_i\varphi_i + b_j\varphi_j + b_m\varphi_m}{2\Delta},\\ \alpha_3 &= \frac{c_i\varphi_i + c_j\varphi_j + c_m\varphi_m}{2\Delta},\end{aligned} \tag{17.5}$$

where

$$\begin{aligned}&a_i = x_j y_m - x_m y_j, \quad &&a_j = x_m y_i - x_i y_m, \quad &&a_m = x_i y_j - x_j y_i,\\ &b_i = y_j - y_m, &&b_j = y_m - y_i, &&b_m = y_i - y_j,\\ &c_i = x_m - x_j, &&c_j = x_i - x_m, &&c_m = x_j - x_i,\end{aligned} \tag{17.6}$$

and

$$2\Delta = \begin{vmatrix} 1 & x_i & y_i \\ 1 & x_j & y_j \\ 1 & x_m & y_m \end{vmatrix}. \tag{17.7}$$

The stationary function φ is approximated by (17.3) so that α_1, α_2, and α_3 in each triangle [or equivalently, ϕ_i, ϕ_j, and ϕ_m, through (17.5)] are taken to minimize (17.2) with respect to φ_i, φ_j, φ_m, φ_r, φ_s, $\varphi_t, \ldots$.

When (17.3) is substituted into (17.2), we have

$$I(\varphi_i, \varphi_j, \varphi_m, \ldots) = \sum \left[\frac{1}{2} \iint_{\text{triangle}} (\alpha_2^2 + \alpha_3^2)\, dx\, dy - \iint_{\text{triangle}} C(\alpha_1 + \alpha_2 x + \alpha_3 y)\, dx\, dy \right]. \tag{17.8}$$

The integrations are defined in the triangle, and Σ means the summation over all triangles. We have

$$\iint_{\text{triangle}} \left(\alpha_2^2 + \alpha_3^2\right) dx\, dy = \left(\alpha_2^2 + \alpha_3^2\right)\Delta,$$

$$\iint_{\text{triangle}} C(\alpha_1 + \alpha_2 x + \alpha_3 y)\, dx\, dy = C(\alpha_1 + \alpha_2 x_0 + \alpha_3 y_0)\Delta \tag{17.9}$$

where C = constant, and x_0 and y_0 are the coordinates at the centroid of the triangle ijm.

When the nodal points are on the boundary Γ, φ must take the given value φ_0.

The unknowns φ_i, φ_j, φ_m, φ_s, φ_t,... are determined from the stationary conditions

$$\frac{\partial I}{\partial \varphi_i} = 0, \qquad \frac{\partial I}{\partial \varphi_j} = 0, \; \ldots. \tag{17.10}$$

It should be noted that φ_i, φ_j, and φ_m appear not only in the shaded triangle in Fig. 17.1 but also in the neighboring triangles.

Example 17.1

An infinite plate with a circular hole of radius a is subjected to uniaxial tension p; see Fig. 17.2. Find t!. , flow of stress around the hole by the finite element method. The domain near the hole is

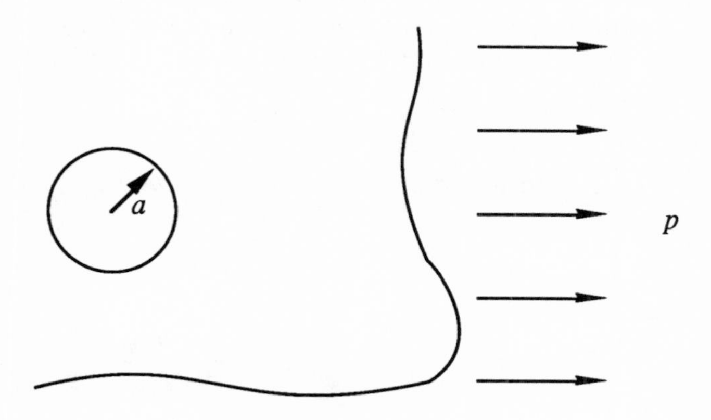

Figure 17.2. A hole in an infinite plate.

divided into the elements shown in Fig. 17.3. In each element, the displacement components u and v are assumed to be linear polynomials,

$$\begin{aligned} u &= \alpha_1 + \alpha_2 x + \alpha_3 y, \\ v &= \alpha_4 + \alpha_5 x + \alpha_6 y. \end{aligned} \tag{17.11}$$

The nodal displacements at points i, j, and m are

$$\begin{aligned} u_i &= \alpha_1 + \alpha_2 x_i + \alpha_3 y_i, \\ v_i &= \alpha_4 + \alpha_5 x_i + \alpha_6 y_i, \\ u_j &= \alpha_1 + \alpha_2 x_j + \alpha_3 y_j, \\ v_j &= \alpha_4 + \alpha_5 x_j + \alpha_6 y_j, \\ u_m &= \alpha_1 + \alpha_2 x_m + \alpha_3 y_m, \\ v_m &= \alpha_4 + \alpha_5 x_m + \alpha_6 y_m. \end{aligned} \tag{17.12}$$

$\alpha_1, \alpha_2, \ldots, \alpha_6$ are solved from (17.12) in terms of the nodal values and the coordinates and then substituted into (17.11), after which we have

$$\begin{aligned} u &= \frac{1}{2\Delta}\left[(a_i + b_i x + c_i y)u_i + (a_j + b_j x + c_j y)u_j + (a_m + b_m x + c_m y)u_m\right], \\ v &= \frac{1}{2\Delta}\left[(a_i + b_i x + c_i y)v_i + (a_j + b_j x + c_j y)v_j + (a_m + b_m x + c_m y)v_m\right], \end{aligned} \tag{17.13}$$

where $a_i, \ldots, \Delta$ are defined by (17.6) and (17.7). Although the given domain is infinite, the triangularization is done for a finite domain. Because of symmetry, a calculation is performed for the first quadrant, u_i is zero at $x = 0$, and v_i is zero at $y = 0$. u and v expressed by (17.13) are substituted into

$$I[u,v] = \frac{1}{2}\int_0^A \int_0^B \left(\sigma_x \varepsilon_x + \sigma_y \varepsilon_y + \tau_{xy}\gamma_{xy}\right) dx\,dy - \int_0^B p u\, dy, \tag{17.14}$$

where

$$\begin{aligned} \varepsilon_x &= \frac{\partial u}{\partial x}, & \varepsilon_y &= \frac{\partial v}{\partial y}, & \gamma_{xy} &= \left(\frac{\partial u}{\partial y} + \frac{\partial v}{\partial x}\right), \\ \sigma_x &= \frac{E}{1-\nu^2}\left(\varepsilon_x + \nu\varepsilon_y\right), & \sigma_y &= \frac{E}{1-\nu^2}\left(\varepsilon_y + \nu\varepsilon_x\right), & \tau_{xy} &= \frac{E}{2(1+\nu)}\gamma_{xy}. \end{aligned} \tag{17.15}$$

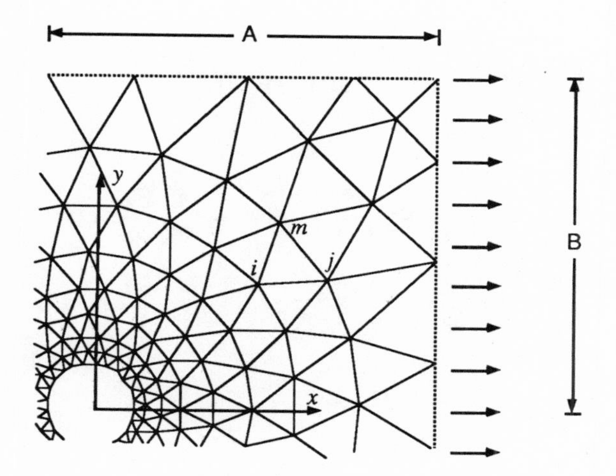

Figure 17.3. Finite element mesh.

Then the functional (17.14) becomes a function of u_i, v_i,.... The stationary conditions are

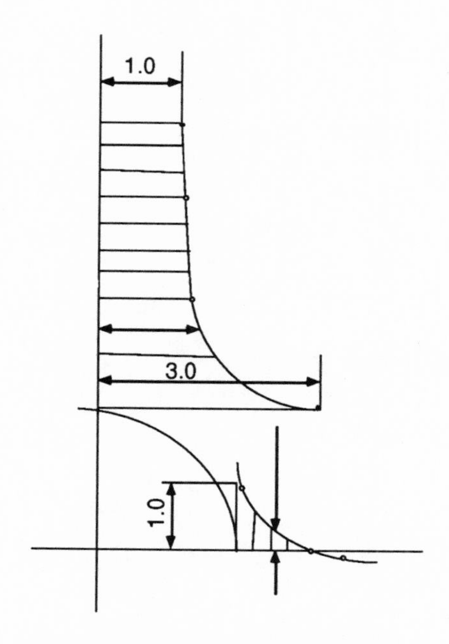

Figure 17.4. Numerical results.

$$\delta I = \int_0^A \int_0^B \left(\sigma_x \delta\varepsilon_x + \sigma_y \delta\varepsilon_y + \tau_{xy}\delta\gamma_{xy}\right) dx\,dy - \int_0^B p\,\delta u\,dy = 0 \qquad (17.16)$$

or

$$\begin{aligned} &\int_0^A \int_0^B \left(\sigma_x \frac{\partial \varepsilon_x}{\partial u_i} + \sigma_y \frac{\partial \varepsilon_y}{\partial u_i} + \tau_{xy} \frac{\partial \gamma_{xy}}{\partial u_i}\right) dx\,dy - \int_0^B p \frac{\partial u}{\partial u_i} dy = 0, \\ &\int_0^A \int_0^B \left(\sigma_x \frac{\partial \varepsilon_x}{\partial v_i} + \sigma_y \frac{\partial \varepsilon_y}{\partial v_i} + \tau_{xy} \frac{\partial \gamma_{xy}}{\partial v_i}\right) dx\,dy - \int_0^B p \frac{\partial u}{\partial v_i} dy = 0, \\ &\ldots, \\ &\ldots. \end{aligned} \qquad (17.17)$$

The number of equations in (17.17) is equal to the number of unknowns u_i, v_i,.... The integrations in (17.17) are the sum of the integrations in the triangular elements.

Numerical results obtained by O. C. Zienkiewicz, Y. K. Cheung, and K. G. Stagg (*J. Strain Analysis 1*, 172-182,1966) are shown in Fig. 17.4 when $p = 1$. The solid curves are the exact solution, and the small circle points present the finite element solution.

18

General Use of the Lagrange Multiplier

Example 18.1.
This example has been given by L. Rosenberg and others (E. Gerjuoy, A. R. P. Rau, L. Rosenberg, and L. Spruch, "Useful extremum principle for the variational calculation of matrix elements," *Phys. Rev.* A9, pp. 108–117, 1974).

Suppose we try to find the root of an equation

$$f(x) = 0. \tag{18.1}$$

In order to do that, we consider the function $I(x)$

$$I(x) = x - \lambda f(x), \tag{18.2}$$

where λ is the Lagrange multiplier. The value of I becomes the root if x satisfies (18.1). Namely, a stationary value of I gives the root. Thus,

$$\delta I = \delta x - \lambda f'(x)\delta x = 0. \tag{18.3}$$

Then

$$\lambda = \frac{1}{f'(x)}. \tag{18.4}$$

Then

$$I(x) = x - \frac{f(x)}{f'(x)} \tag{18.5}$$

gives an approximate value for the root by choosing an arbitrary value of x. The approximation becomes better when x is close to the exact solution. This result is nothing but the **Newton method**.

The usage of the Lagrange multiplier is a little different in this section because λ is fixed when the variation of I is taken.

Example 18.2.
Another elementary example is shown. Suppose we try to compute a value of $y(x)$ at $x = 1$, where $y(x)$ is the solution of the differential equation

$$y' + y^2 = 0, \qquad 0 < x < 1 \tag{18.6}$$

with the boundary condition $y(0) = 1$. Consider the functional

$$I[y] = y(1) - \int_0^1 \lambda\left(y' + y^2\right) dx. \tag{18.7}$$

This functional is formulated so that its stationary value becomes $y(1)$. In other words, $I[y]$ becomes $y(1)$ when (18.6) is satisfied. Equation (18.6) is a stationary condition when the variation of λ is taken. When λ is fixed, the stationary conditions are obtained by

$$\begin{aligned} \delta I &= \delta y(1) - \int_0^1 \lambda(\delta y' + 2y\delta y')' dx \\ &= \delta y(1) - [\lambda \delta y]_0^1 + \int_0^1 (\lambda' - 2y\lambda)\delta y\, dx = 0. \end{aligned} \tag{18.8}$$

Trial functions must satisfy the boundary conditions $y(0) = 1$ or $\delta y(0) = 0$. Then (18.8) leads to

$$\begin{aligned} 1 - \lambda(1) &= 0, \\ \lambda' - 2y\lambda &= 0. \end{aligned} \tag{18.9}$$

λ to satisfy (18.9) is

$$\lambda = \exp\left(2\int_1^x y\, dx\right). \tag{18.10}$$

An approximate value of I is obtained by choosing a trial function y satisfying the boundary condition $y(0)=1$ and by choosing λ given by (18.10). For instance, we choose the trial function y as a constant 1. Then $\lambda = \exp[2(x - 1)]$. When it is substituted into (18.7),

$$I[y] = 1 - \int_0^1 e^{2(x-1)}\, dx = \frac{1}{2}\left(1 + e^{-2}\right) = \frac{1.135}{2}. \tag{18.11}$$

The true solution is $y = 1/(1 + x)$ and $y(1) = \frac{1}{2}$ which is very close to the value given in (18.11).

Example 18.3.
Consider a standard eigenvalue problem

$$L[u] - \mu u = 0 \tag{18.12}$$

with homogeneous boundary conditions, where L is a given differential operator. Since we want the eigenvalue μ, we formulate the functional

$$I[u] = \mu - \int_G \lambda(L[u] - \mu u)\, d\mathbf{x}, \tag{18.13}$$

where G is a domain in any dimension and $d\mathbf{x}$ is an element of G. The stationary condition for the variation of λ is (18.12) and the stationary value of I is μ. When λ is fixed, the stationary condition is obtained by

$$\begin{aligned} \delta I &= \delta\mu - \int_G \lambda\big(L[\delta u] - \delta\mu u - \mu\delta u\big)\, d\mathbf{x} \\ &= \delta\mu - \int_G \big[L(\lambda) - \mu\lambda\big]\delta u\, d\mathbf{x} + \delta\mu \int_G \lambda u\, d\mathbf{x}, \end{aligned} \tag{18.14}$$

and, therefore,

$$\begin{aligned} &1 + \int_G \lambda u\, d\mathbf{x} = 0, \\ &L(\lambda) - \mu\lambda = 0. \end{aligned} \tag{18.15}$$

We can see that

$$\lambda = -\frac{u}{\int_G u^2\, d\mathbf{x}} \tag{18.16}$$

satisfies the conditions (18.15) because

$$L(\lambda) = -\frac{L[\lambda]}{\int_G u^2\, d\mathbf{x}} = -\frac{\mu u}{\int_G u^2\, d\mathbf{x}} = \mu\lambda. \tag{18.17}$$

When (18.16) is substituted into (18.13), we have

$$I[u] = \frac{\int_G uL[u]\,d\mathbf{x}}{\int_G u^2\,d\mathbf{x}}, \tag{18.18}$$

which is nothing but the **Rayleigh equation.**

A value of I with a trial function u gives an approximate value of the eigenvalue.

Example 18.4.

A perfectly plastic continuum G is subjected to a given surface force F_i on S. The problem is to find the **safety factor** m. The material starts collapsing plastically by the force mF_i At the state of plastic collapse, the stress state must satisfy the conditions

$$\begin{aligned} \left(s_{ij} + \delta_{ij}\sigma\right)_{,j} &= 0 \qquad \text{in } G, \\ \left(s_{ij} + \delta_{ij}\sigma\right)n_j &= mF_i \qquad \text{in } S, \\ f\left(s_{ij}\right) \le 0 \quad \text{or} \quad & f\left(s_{ij}\right) + \varphi^2 = 0, \end{aligned} \tag{18.19}$$

where s_{ij} is the stress deviatoric, σ the mean hydrostatic stress, n_j the normal vector on surface S, and $f(s_{ij})$ the yielding condition

$$f\left(s_{ij}\right) = \tfrac{1}{2} s_{ij} s_{ij} - k^2. \tag{18.20}$$

T. Mura and S. L. Lee ("Application of variational principles to limit analysis," *Q. J. Appl. Math.* 21, pp. 243–248, 1963) constructed the following functional that leads to the safety factor at the stationary state:

$$\begin{aligned} I\left[m, s_{ij}, \sigma, \varphi, v_i, \mu\right] = m &- \int_G v_i\left(s_{ij} + \delta_{ij}\sigma\right)_{,j} d\mathbf{x} \\ &- \int_S \lambda_i\left[mF_i - \left(s_{ij} + \delta_{ij}\sigma\right)n_j\right] dS \\ &- \int_G \mu\left[f\left(s_{ij}\right) + \varphi^2\right] d\mathbf{x}, \end{aligned} \tag{18.21}$$

where v_i, λ_i and μ are the Lagrange multipliers and φ is introduced to take care of the inequality condition in (18.19). The stationary conditions for v_i, λ_i, μ and φ are obtained by $\delta I = 0$,

$$\delta I = \delta m - \int_G v_i\left(\delta s_{ij} + \delta_{ij}\delta\sigma\right)_{,j} d\mathbf{x} - \int_S \lambda_i\left[\delta m F_i - \left(\delta s_{ij} + \delta_{ij}\delta\sigma\right)n_j\right] dS$$
$$- \int_G \mu\left(\frac{\partial f}{\partial s_{ij}}\delta s_{ij} + 2\varphi\delta\varphi\right) d\mathbf{x} = 0. \tag{18.22}$$

Then we have

$$\begin{aligned} 1 &= \int_S \lambda_i F_i \, dS, & \lambda_i &= v_i & &\text{on } S, \\ \frac{1}{2}\left(v_{i,j} + v_{j,i}\right) &= \mu \frac{\partial f}{\partial s_{ij}} & \mu\varphi &= 0 & &\text{in } G, \\ v_{i,j}\delta_{ij} &= 0 & & & &\text{in } G. \end{aligned} \tag{18.23}$$

In addition to conditions (18.19), conditions (18.23) are stationary conditions for the collapse state. It is obvious that v_i is the velocity vector.

The first condition in (18.23), $\int_S \lambda_i F_i \, dS = 1$, means that the work done by F_i is normalized for comparison. The second condition means that the strain rate $\dot{\varepsilon}_{ij}$ is proportional to $\partial f / \partial s_{ij} = s_{ij}$ when $\varphi = 0$, and that $\dot{\varepsilon}_{ij} = 0$ when $\varphi \neq 0$. The case $\varphi = 0$ corresponds to $\frac{1}{2}s_{ij}s_{ij} = k^2$, and the case $\varphi \neq 0$ corresponds to $\frac{1}{2}s_{ij}s_{ij} \leq k^2$, i.e., the domain where $\varphi = 0$ is in a plastic state and the domain where $\varphi \neq 0$ is still in an elastic state. This indicates also that the elastic strain does not change during the collapse. The last condition is the condition for no dilatation (no volume change). This is true because the strain change during the collapse is only plastic.

When conditions (18.23) are used, the functional (18.21) becomes

$$I\left[\mu, s_{ij}\right] = \tfrac{1}{2}\int_G \mu s_{ij}s_{ij} \, d\mathbf{x} + \int_G \mu k^2 \, d\mathbf{x}. \tag{18.24}$$

The constraint conditions for this variation are (18.19).

The tension specimen subjected to uniformly distributed tensile stress mp per unit length, shown in Figs. 18.1*a*, *b*, is used to illustrate the application of the theorems, assuming a state of plane stress. Due to symmetry, only one-quarter of the specimen needs to be considered.

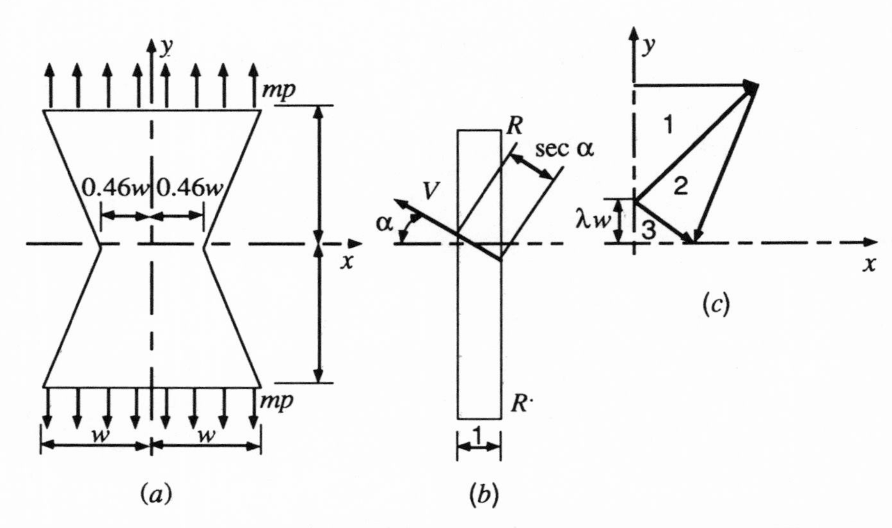

Figure 18.1. (a) Tension specimen subjected to uniformly distributed tensile stress, (b) Side view of the specimen, (c) Assumed stress field.

The assumed stress field consists of zones of constant stress fields separated by lines of discontinuity, as shown in Fig. 18.1*c*. T. Mura, W. H. Rimawi, and S. L. Lee ("Extended theorems of limit analysis," *Q.J. Appl. Math.* 23, pp. 171–179, 1965) have found $I \approx 0.911k/p$ as an approximation of the safety factor. The true value of the safety factor m is between $0.911k/p$ and $0.920k/p$.

19

Miscellaneous Problems

Variational principles take different forms in fracture mechanics and micromechanics.

Example 19.1

Find a critical crack extension under the given applied stress σ_{22}^0 at infinity in a infinite body for a slitlike crack in the x axis with length $2a$ (see Fig. 19.1).

If there is no crack, the applied stress σ_{22}^0 at infinity is uniform everywhere. If there is a slitlike crack along the x axis, the stress field is disturbed and the resulting stress is $\sigma_{ij}^0 + \sigma_{ij}$ in the neighborhood of the crack, where $\sigma_{ij}^0 = \sigma_{22}^0$ and the other components are zero. The stress disturbance σ_{22} is simulated by an **eigenstrain** (nonelastic strain or fictitious transformation strain) ε_{22}^* distributed along the domain $|x| \le a$, in the homogeneous material. Then the stress σ_{22} in $|x| \le a$, $y = 0$ is obtained as

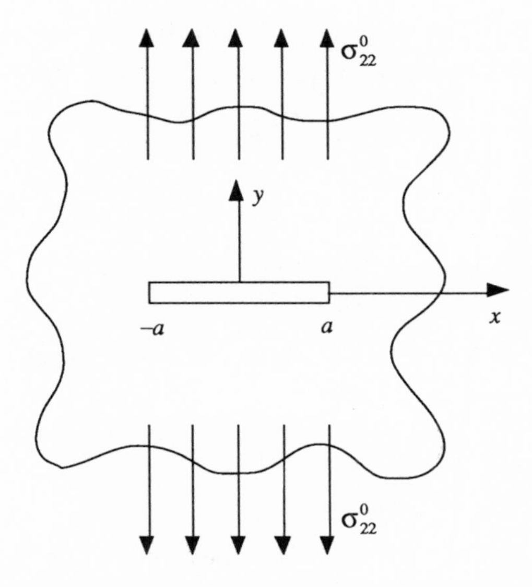

Figure 19.1. Crack with length $2a$ subjected to uniform tension at infinity.

$$\sigma_{22} = -\frac{\mu}{1-\nu}\frac{\varepsilon^*}{a}, \tag{19.1}$$

where μ is the shear modulus, ν is Poisson's ratio, and $\varepsilon^* = \varepsilon^*_{22} a_2$ $(a_2 \to 0,\ \varepsilon^*_{22} \to \infty)$, a_2 being the minor semiaxis of the elliptical crack (see T. Mura, *Micromechanics of Defects in Solids,* p. 206, Martinus Nijhoff, 1982). The value ε^* is determined by the boundary condition

$$\sigma^0_{22} + \sigma_{22} = 0 \quad \text{at } y = 0,\ |x| < a. \tag{19.2}$$

Then we have

$$\varepsilon^* = (1-\nu)a\frac{\sigma^0_{22}}{\mu}. \tag{19.3}$$

The interaction energy between the applied stress σ_{ij} and ε^*_{ij} is written as

$$\Delta W = -\frac{1}{2}\iiint \sigma^0_{ij}\varepsilon^*_{ij}\, dv. \tag{19.4}$$

The domain where ε^*_{ij} is distributed is the infinitely thin elliptical domain in a unit thickness of the plate. Then it becomes

$$\begin{aligned} \Delta W &= -\tfrac{1}{2}\sigma^0_{22}\varepsilon^*_{22}\pi a a_2 \\ &= -\tfrac{1}{2}\sigma^0_{22}\pi a \varepsilon^* \\ &= -\pi(1-\nu)a^2\frac{\left(\sigma^0_{22}\right)^2}{2\mu}. \end{aligned} \tag{19.5}$$

The **Griffith fracture criteria** is obtained by the stationary condition,

$$\frac{\partial}{\partial a}(\Delta W + 4a\gamma) = 0, \tag{19.6}$$

where γ is the surface energy. The above equation leads to

$$\sigma^0_{22} = \sqrt{\frac{4\mu\gamma}{\pi(1-\nu)a}}. \tag{19.7}$$

The crack of length $2a$ becomes unstable for an applied stress larger than the value given by (19.7). Or cracks with length larger than the value given by (19.7) under a given stress σ_{22}^0 become unstable.

Example 19.2
A material contains hard particles with volume fraction f as shown in Fig. 19.2. Young's modulus and Poisson's ratio of the matrix are E and ν, and those for the particles are E^* and ν^*. The matrix is perfectly elastoplastic with the yield stress σ_y, and the particles are perfectly elastic. Find the uniaxial stress and strain curve of this composite material.

Let us denote the applied stress by σ_{33}^0 at infinity. We assume that the plastic strain in the matrix is uniform and that $\varepsilon_{33}^P = \varepsilon^P$ and $\varepsilon_{11}^P = \varepsilon_{22}^P = -\varepsilon^P/2$ because the nondilatation condition of the plastic strain requires $\varepsilon_{11}^P + \varepsilon_{22}^P + \varepsilon_{33}^P = 0$. The condition for yielding is

$$\sigma_{ij}^0 + \sigma_{ij} = \sigma_y \tag{19.8}$$

in the matrix, where σ_{ij} is the stress disturbance due to the particles and the plastic strain ε_{ij}^P in the matrix (misfit strain). Galerkin's method leads to

$$\iiint_D \left(\sigma_{ij}^0 + \sigma_{ij} - \sigma_y\right)\delta\varepsilon_{ij}^P \, dv = 0. \tag{19.9}$$

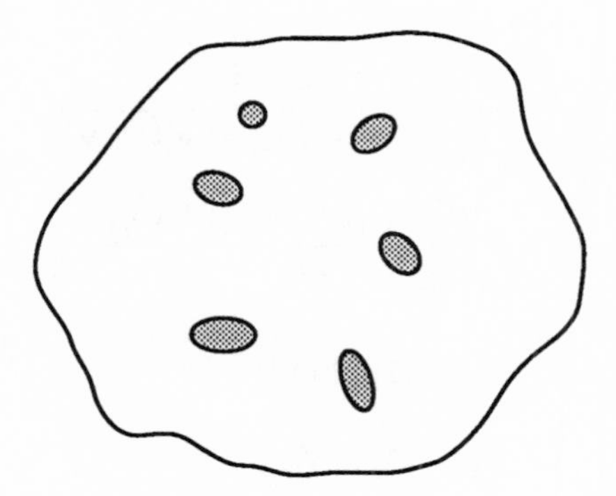

Figure 19.2. Perfectly elastoplastic material containing hard particles.

The domain where $\delta\varepsilon_{ij}^P$ is defined is limited to the matrix, $\sigma_{ij}^0 = \sigma_{33}^0$, and σ_y has only components in the z direction; then (19.9) is written as

$$\iiint\limits_{D-\Omega} \left(\sigma_{33}^0 \delta\varepsilon_{33}^P + \sigma_{11}\delta\varepsilon_{11}^P + \sigma_{22}\delta\varepsilon_{22}^P + \sigma_{33}\delta\varepsilon_{33}^P - \sigma_y \delta\varepsilon_{33}^P\right) dv = 0, \tag{19.10}$$

where Ω is the total volume (domain) of the particles. Since $\delta\varepsilon_{33}^P = \delta\varepsilon^P$, $\delta\varepsilon_{11}^P = \delta\varepsilon_{22}^P = -\delta\varepsilon^P/2$, and they are uniform in $D-\Omega$, we can write (19.10) as

$$\iiint\limits_{D-\Omega} \left(\sigma_{33}^0 - \frac{\sigma_{11}}{2} - \frac{\sigma_{22}}{2} + \sigma_{33} - \sigma_y\right) \delta\varepsilon^P dv = 0 \tag{19.11}$$

or

$$\left(\sigma_{33}^0 - \sigma_y\right) \iiint\limits_{D-\Omega} dv + \iiint\limits_{D-\Omega} \left(\sigma_{33} - \frac{\sigma_{11}}{2} - \frac{\sigma_{22}}{2}\right) dv = 0.$$

Tanka and Mori (*Acta Met 18*, pp. 931–941, 1970) evaluated the mean values of σ_{33}, σ_{11}, and σ_{22} in the matrix as follows:

$$\iiint\limits_{D-\Omega} \frac{\left(\sigma_{33} - \frac{\sigma_{11}}{2} - \frac{\sigma_{22}}{2}\right) dv}{\iiint\limits_{D-\Omega} dv} = -f\left(A\varepsilon_{33}^P + B\sigma_{33}^0\right), \tag{19.12}$$

where

$$A = \frac{\frac{3}{2} E^* E(7 - 5\nu)}{E^*(1+\nu)(8 - 10\nu) + E(7 - 5\nu)(1 + \nu^*)},$$
$$B = \frac{(7 - 5\nu)\left[E^*(1+\nu) - E(1+\nu^*)\right]}{E^*(1+\nu)(8 - 10\nu) + E(7 - 5\nu)(1 + \nu^*)}, \tag{19.13}$$

for spherical particles. Then (19.11) leads to

$$\sigma_{33}^0 = \frac{\sigma_y + fA\varepsilon_{33}^P}{1 - fB}. \tag{19.14}$$

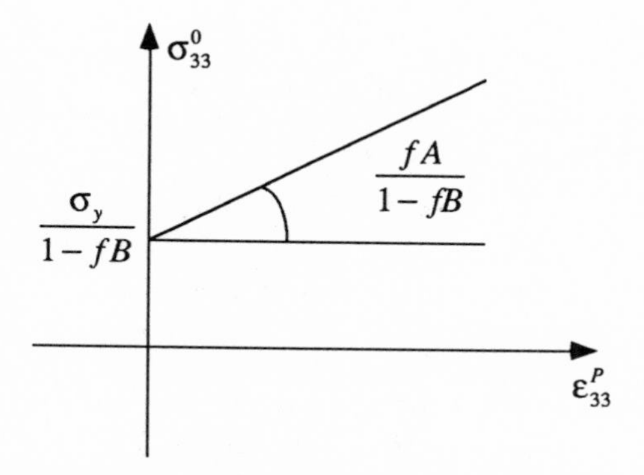

Figure 19.3. Dispersion hardening by particles.

The stress or plastic strain curve is shown in Fig. 19.3. This is the Tanaka-Mori theory on linear work hardening. The dispersion of hard particles increases the yield stress and makes the perfectly plastic material a hardening material.

Example 19.3
Ingot iron tensile specimens that were treated by strain aging at 303K during 860 seconds show different stress-strain curves at testing temperatures of 303K and 77K, as shown in Fig. 19.4. The tensile test at 303K shows a clear yield point while the test at 77K

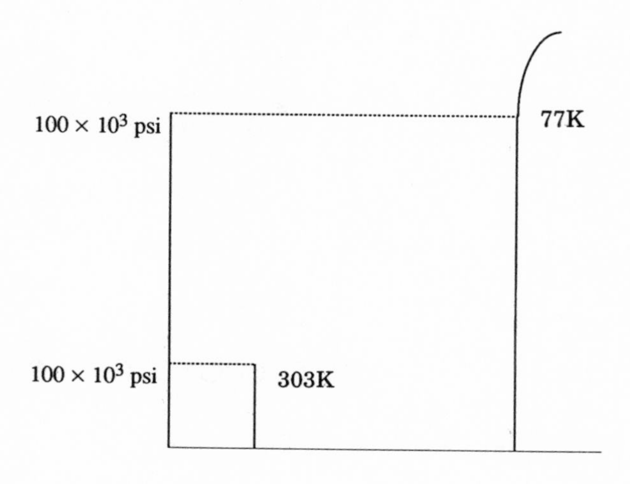

Figure 19.4. Stress-strain diagrams of carbon iron at temperature 303K and 77K.

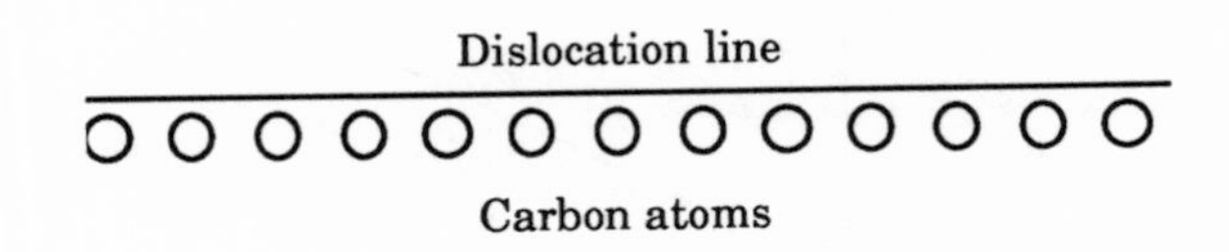

Figure 19.5. Carbon atoms piping along a straight dislocation.

does not. Explain this phenomenon by the interaction between dislocations and carbon atoms (see Mura and Brittain, *Acta Met. 8*, p. 767, 1960).

During strain aging, carbon atoms diffuse to dislocations according to the **Cottrell and Bilby theory.** After aging, a dislocation-line configuration and an array of carbon atoms become as shown in Fig. 19.5.

When the stress σ is applied in the tensile test, the dislocation line deforms as shown in Fig. 19.6. Some points on the dislocation line are still locked in by the carbon atoms.

Perhaps the interval ℓ between the two pinpoints depends on the testing temperature; a higher testing temperature provides a large ℓ. Let us calculate the associate energies caused by the dislocation displacement between $x = 0$ and $x = 1$. The increase in the potential energy of the dislocation due to line tension T is

$$\int_0^{\ell} \frac{1}{2} (y')^2 T \, dx. \tag{19.15}$$

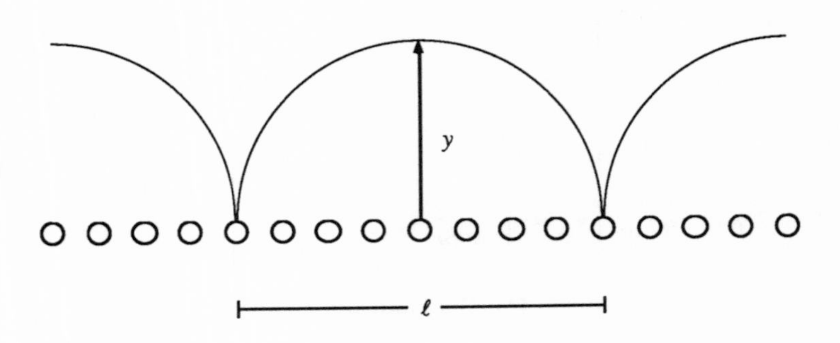

Figure 19.6. Dislocation line escapes from the carbon array by applied force.

The interaction energy between the dislocation and the array of carbon atoms is approximately

$$Ac\int_0^\ell \frac{p}{y^2+p^2}\,dx, \tag{19.16}$$

where A is the interaction constant about $3\times10^{-21}\,\mathrm{dyn/m^2}$, p is the distance between the dislocation line and the array of carbon atoms at the initial configuration if Fig. 19.5 is measured normal to the plane of the paper, and c is the number of carbon atoms per unit length of the array. The work done by applied stress is

$$\int_0^\ell \sigma b y\,dx, \tag{19.17}$$

where b is the **Burgers vector** (magnitude of slip). The Gibbs free energy is

$$I[y]=\int_0^\ell \frac{T}{2}(y')^2\,dx + Ac\int_0^\ell \frac{p}{y^2+p^2}\,dx - \int_0^\ell \sigma b y\,dx \tag{19.18}$$

with the boundary conditions $y=0$ at $x=0$ and ℓ. The Euler equation of $I[y]$ leads to

$$\sigma b + y''T - Acp\frac{2y}{\left(y^2+p^2\right)^2}=0. \tag{19.19}$$

The first integration of (19.19) is obtained by multiplying by y',

$$y\sigma b + \frac{T}{2}(y')^2 + \frac{Acp}{y^2+p^2}=C_1. \tag{19.20}$$

Further integration is obtained as

$$x=\sqrt{\frac{T}{2}}\int_0^y \frac{1}{\sqrt{C_1 - y\sigma b - Acp/\left(y^2+p^2\right)}}\,dy. \tag{19.21}$$

The integral constant C_1 is determined by $y=0$ and $x=\ell$. The relationship between σ and the maximum deflection $y(\ell/2)$ is obtained from (19.21). It is shown with parameter λ, where

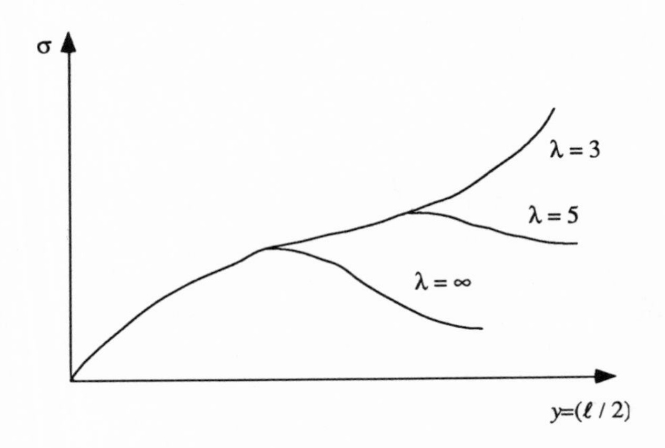

Figure 19.7. Applied force vs. dislocation displacements. Appearance of the maximum corresponds to yielding.

$$\lambda = \sqrt{\frac{\ell^2 Ac}{2Tp^3}}. \tag{19.22}$$

According to the Cottrell theory, the yield point appears when the σ-y curve takes a maximum point. Figure 19.7 shows that low values of λ do not give the maximum point. Therefore, it is concluded by this calculation that when λ is small, no yield point is expected. When the testing temperature is low, a small ℓ is expected because no sufficient thermal energy is supplied to break the locking (pin) points. Perhaps ℓ^2/T is an increasing function of T so that large values of λ are obtained for large T.

20

More Numerical Methods

In Chapters 3 and 8, we examined Ritz's and Galerkin's methods for solving differential equations. In Chapter 18 we have derived Newton's method to solve 1-D nonlinear equations. In this chapter we apply some of these methods to different problems and also investigate other techniques derivable from the variational method point of view.

Solution of Linear Algebraic Equations
Example 20.1

Consider the following system of simultaneous linear equations:

$$\begin{bmatrix} 1 & 1 \\ 3 & 4 \\ 2 & 2 \end{bmatrix} \begin{Bmatrix} x_1 \\ x_2 \end{Bmatrix} = \begin{Bmatrix} 2 \\ 7 \\ 5 \end{Bmatrix}. \tag{20.1}$$

This is a very irregular problem because the coefficient matrix of (20.1) is not square; consequently, we cannot apply the well-known Gauss elimination technique. We also notice that the first and the third equations are linearly dependent. This type of problem is called an **ill-posed** problem.

In most cases, we cannot expect to obtain the exact solution from a problem like (20.1); however, we may be able to obtain a close solution in a certain sense. For instance, let us denote a close solution as $\{\bar{x}_1 \quad \bar{x}_2\}^{\mathrm{T}}$ and substitute it into (20.1) and subtract the result from the right-hand side. We have

$$\boldsymbol{\varepsilon} = \begin{Bmatrix} 2 \\ 7 \\ 5 \end{Bmatrix} - \begin{Bmatrix} \bar{x}_1 + \bar{x}_2 \\ 3\bar{x}_1 + 4\bar{x}_2 \\ 2\bar{x}_1 + 2\bar{x}_2 \end{Bmatrix} = \begin{Bmatrix} 2 - \bar{x}_1 - \bar{x}_2 \\ 7 - 3\bar{x}_1 - 4\bar{x}_2 \\ 5 - 2\bar{x}_1 - 2\bar{x}_2 \end{Bmatrix} \tag{20.2}$$

where $\boldsymbol{\varepsilon}$ is a vector which is not necessarily a zero vector. Suppose we look for a solution that minimizes the squared magnitude of $\boldsymbol{\varepsilon}$. We can write this problem as

$$\text{minimize } \|\mathbf{Ax} - \mathbf{b}\|^2 \tag{20.3}$$

where $\|\cdot\|$ is the N dimensional Euclidean norm

$$\|\mathbf{v}\| = \sqrt{\sum_{i=1}^{N} v_i^2}$$

and

$$\mathbf{A} = \begin{bmatrix} 1 & 1 \\ 3 & 4 \\ 2 & 2 \end{bmatrix}, \quad \mathbf{x} = \{\bar{x}_1 \quad \bar{x}_2\}^{\mathrm{T}}, \quad \mathbf{b} = \{2 \quad 7 \quad 5\}^{\mathrm{T}}. \tag{20.4}$$

When (20.3) is differentiated with respect to $\mathbf{x}$ and set to zero, we have

$$\mathbf{A}^{\mathrm{T}}\mathbf{A}\mathbf{x} = \mathbf{A}^{\mathrm{T}}\mathbf{b}, \tag{20.5}$$

and (20.5) is called the **normal equation.** Its solution is

$$\mathbf{x} = \left(\mathbf{A}^{\mathrm{T}}\mathbf{A}\right)^{-1}\mathbf{A}^{\mathrm{T}}\mathbf{b}. \tag{20.6}$$

The matrix $\left(\mathbf{A}^{\mathrm{T}}\mathbf{A}\right)^{-1}\mathbf{A}^{\mathrm{T}}$ is called the **general inverse** of $\mathbf{A}$.

Let us apply (20.6) to the current problem.

$$\mathbf{x} = \left(\begin{bmatrix} 1 & 3 & 2 \\ 1 & 4 & 2 \end{bmatrix} \begin{bmatrix} 1 & 1 \\ 3 & 4 \\ 2 & 2 \end{bmatrix} \right)^{-1} \begin{bmatrix} 1 & 3 & 2 \\ 1 & 4 & 2 \end{bmatrix} \begin{Bmatrix} 2 \\ 7 \\ 5 \end{Bmatrix} = -\frac{1}{5} \begin{Bmatrix} 13 \\ -1 \end{Bmatrix}. \tag{20.7}$$

We can check the accuracy by substituting $\mathbf{x}$ into (20.1).

$$\frac{1}{5} \begin{bmatrix} 1 & 1 \\ 3 & 4 \\ 2 & 2 \end{bmatrix} \begin{Bmatrix} 13 \\ -1 \end{Bmatrix} = \frac{1}{5} \begin{Bmatrix} 12 \\ 35 \\ 24 \end{Bmatrix} = \begin{Bmatrix} 2.4 \\ 7 \\ 4.8 \end{Bmatrix}. \tag{20.8}$$

We notice that whereas the second equation is satisfied exactly, the first and the third equations are satisfied approximately. This method is called the **least-squares method** as seen in Chapter 3.

We notice that this method works as long as the inverse operation in (20.6) is valid. Because $\mathbf{A}^T\mathbf{A}$ in (20.7) is a 2×2 matrix, the rank of the matrix must be 2 in order for it to have the inverse. In this example problem, the inverse operation was valid because the rank of the matrix **A** was 2; in other words, for a $m\times n$ matrix **A** to have a least-squares solution, the rank of **A** must be n.

Example 20.2

Consider

$$\begin{bmatrix} 2 & -1 & 0 \\ -1 & 2 & -1 \\ 0 & -1 & 2 \end{bmatrix} \begin{Bmatrix} x_1 \\ x_2 \\ x_3 \end{Bmatrix} = \begin{Bmatrix} -1 \\ 0 \\ 7 \end{Bmatrix}. \tag{20.9}$$

When the coefficient matrix is symmetric and positive definite (i.e., the eigenvalues are all positive) as in this example, the quadratic function

$$I[\mathbf{x}] = \tfrac{1}{2}\mathbf{x}\mathbf{A}\mathbf{x} - \mathbf{b}\mathbf{x}, \tag{20.10}$$

where

$$\begin{aligned} \mathbf{A} &= \begin{bmatrix} 2 & -1 & 0 \\ -1 & 2 & -1 \\ 0 & -1 & 2 \end{bmatrix}, \\ \mathbf{x} &= \{x_1 \quad x_2 \quad x_3\}^T, \\ \mathbf{b} &= \{-1 \quad 0 \quad 7\}^T, \end{aligned} \tag{20.11}$$

is known to have a minimum value because (20.10) is analogous to a functional form. The corresponding equation analogous to the Euler equation of (20.10) is

$$\mathbf{A}\mathbf{x} = \mathbf{b} \tag{20.12}$$

which is the original problem. Therefore, we conclude that the solution that minimizes (20.10) is the solution of the original problem (20.9).

We are going to solve this problem by iteration, that is we first guess a solution and refine the solution progressively. Let us denote our initial guess as $\mathbf{x}_0$ and the successive refined solutions by $\mathbf{x}_k$ $(k=1,2,3,\ldots)$. Suppose the next solution and the current solution are related by

$$\mathbf{x}_{k+1} = \mathbf{x}_k + \alpha_k \mathbf{s}_k \tag{20.13}$$

where $\mathbf{s}_k$ is the kth search direction vector, and α_k is the kth step length scalar. Instinctively, we want to move towards the direction in which I decreases. This means we must proceed towards the negative direction of the gradient of I. Then the search direction is

$$-\nabla I = -\mathbf{A}\mathbf{x} + \mathbf{b} = \mathbf{r} = \mathbf{s} \tag{20.14}$$

where the vector $\mathbf{r}$ is the residual error due to applying an approximate solution for $\mathbf{x}$. Let us substitute $\mathbf{x}_0$ into (20.14). Since it is very unlikely our guess is perfect, (20.14) yields a nonzero vector $\mathbf{r}_0$, namely, $\mathbf{s}_0$, which becomes our first search direction. Now that we have a search direction, we would like to know how far we can proceed in the direction; in other words, we would like to know α_0. Because we hope to achieve eventually $-\nabla I(\mathbf{x}_k + \alpha_k \mathbf{s}_k) = 0$, we can compute α_0 from (20.14) by

$$\begin{aligned} \mathbf{A}(\mathbf{x}_0 + \alpha_0 \mathbf{s}_0) &= \mathbf{b} \\ \alpha_0 \mathbf{A}\mathbf{s}_0 &= \mathbf{b} - \mathbf{A}\mathbf{x}_0 = \mathbf{r}_0 = \mathbf{s}_0 \\ \alpha_0 &= \frac{\mathbf{s}_0^{\mathrm{T}} \mathbf{s}_0}{\mathbf{s}_0^{\mathrm{T}} \mathbf{A}\mathbf{s}_0}. \end{aligned} \tag{20.15}$$

Then we have the next solution $\mathbf{x}_1 = \mathbf{x}_0 + \alpha_0 \mathbf{s}_0$. We can compute $\mathbf{r}_1$ and repeat (20.15). This procedure is called the **method of steepest descent** because we always choose the most negative direction, i.e., the negative of the gradient.

It turns out that choosing the negative gradient direction is not necessarily the best as it is not always the best, for example, to follow the steepest route down the mountain. Let us modify the search direction by

$$\mathbf{s}_{k+1} = \mathbf{r}_{k+1} + \beta_k \mathbf{s}_k \tag{20.16}$$

where β_k is some constant. In addition, we are going to require

$$\mathbf{s}_i^{\mathrm{T}}\mathbf{A}\mathbf{s}_j = 0, \quad i \neq j, \tag{20.17}$$

and $\mathbf{s}_i$ and $\mathbf{s}_j$ are said to be **conjugate** to each other. We can compute β_k from (20.16) and (20.17) as

$$\beta_k = -\frac{\mathbf{r}_{k+1}^{\mathrm{T}}\mathbf{A}\mathbf{s}_k}{\mathbf{s}_k^{\mathrm{T}}\mathbf{A}\mathbf{s}_k}. \tag{20.18}$$

It is interesting to observe from (2.16) and (2.17) that

$$\begin{aligned}
\mathbf{r}_{k+1}^{\mathrm{T}}\mathbf{r}_k &= \left(\mathbf{b} - \mathbf{A}\mathbf{x}_{k+1}\right)^{\mathrm{T}}\mathbf{r}_k \\
&= \left\{\mathbf{b} - \mathbf{A}\left[\mathbf{x}_k + \frac{\mathbf{r}_k^{\mathrm{T}}\mathbf{r}_k}{\mathbf{s}_k^{\mathrm{T}}\mathbf{A}\mathbf{s}_k}\left(\mathbf{r}_k - \frac{\mathbf{r}_k^{\mathrm{T}}\mathbf{A}\mathbf{s}_k}{\mathbf{s}_k^{\mathrm{T}}\mathbf{A}\mathbf{s}_k}\mathbf{s}_k\right)\right]\right\}^{\mathrm{T}}\mathbf{r}_k \\
&= \mathbf{r}_k^{\mathrm{T}}\mathbf{r}_k - \frac{\mathbf{r}_k^{\mathrm{T}}\mathbf{r}_k}{\mathbf{s}_k^{\mathrm{T}}\mathbf{A}\mathbf{s}_k}\mathbf{r}_k^{\mathrm{T}}\mathbf{r}_k + \frac{\mathbf{r}_k^{\mathrm{T}}\mathbf{A}\mathbf{s}_k}{\mathbf{s}_k^{\mathrm{T}}\mathbf{A}\mathbf{s}_k}\mathbf{s}_k^{\mathrm{T}}\mathbf{r}_k \\
&= \mathbf{r}_k^{\mathrm{T}}\mathbf{r}_k - \mathbf{r}_k^{\mathrm{T}}\mathbf{r}_k \\
&= 0
\end{aligned} \tag{20.19}$$

which means that two successive residual vectors are orthogonal. Furthermore, it can be shown that

$$\mathbf{r}_i^{\mathrm{T}}\mathbf{r}_j = 0, \quad i \neq j. \tag{20.20}$$

(20.20) implies that the residual vectors are mutually orthogonal.

In addition, it can be shown that

$$\mathbf{s}_i^{\mathrm{T}}\mathbf{A}\mathbf{s}_j = 0, \quad i \neq j. \tag{20.21}$$

In other words, the search directions are conjugate to one another.

When **A** is a n dimensional matrix, and a vector is orthogonal to n mutually orthogonal residual vectors, the vector must be a zero vector; consequently, the n+1th **r** vector must be a zero vector. When the residual vector is zero then the solution is reached. This method is called the **conjugate gradient method** or the **C G** method for short.

According to the conjugate gradient method, (20.9) can be solved in three iterations. Let us try. First, we choose

$$\mathbf{x}_0 = \{1, \quad 1, \quad 1\}^{\mathrm{T}} \tag{20.22}$$

as our initial guess. The residual vector is

$$\mathbf{r}_0 = \begin{Bmatrix} -1 \\ 0 \\ 7 \end{Bmatrix} - \begin{bmatrix} 2 & -1 & 0 \\ -1 & 2 & -1 \\ 0 & -1 & 2 \end{bmatrix} \begin{Bmatrix} 1 \\ 1 \\ 1 \end{Bmatrix} = \begin{Bmatrix} -2 \\ 0 \\ 6 \end{Bmatrix}. \tag{20.23}$$

Since we have no previous search direction, we take $\mathbf{r}_0$ as the search direction, i.e.,

$$\mathbf{s}_0 = \mathbf{r}_0 = \{-2, \quad 0, \quad 6\}^{\mathrm{T}}. \tag{20.24}$$

The search step length α_0 is

$$\alpha_0 = \frac{\mathbf{r}_0^{\mathrm{T}}\mathbf{r}_0}{\mathbf{s}_0^{\mathrm{T}}\mathbf{A}\mathbf{s}_0} = \frac{\mathbf{r}_0^{\mathrm{T}}\mathbf{r}_0}{\mathbf{r}_0^{\mathrm{T}}\mathbf{A}\mathbf{r}_0} = \frac{40}{80} = \frac{1}{2}. \tag{20.25}$$

Therefore, our next guess is

$$\mathbf{x}_1 = \mathbf{x}_0 + \alpha_0\mathbf{s}_0 = \{0, \quad 1, \quad 4\}^{\mathrm{T}}. \tag{20.26}$$

The next residual vector is

$$\mathbf{r}_1 = \begin{Bmatrix} -1 \\ -2 \\ 7 \end{Bmatrix} - \begin{bmatrix} 2 & -1 & 0 \\ -1 & 2 & -1 \\ 0 & -1 & 2 \end{bmatrix} \begin{Bmatrix} 0 \\ 1 \\ 4 \end{Bmatrix} = \begin{Bmatrix} 0 \\ 2 \\ 0 \end{Bmatrix}. \tag{20.27}$$

Apparently, the residual vector is smaller than before. We now need the next search direction. (20.18) yields

$$\beta_0 = -\frac{\mathbf{r}_1^{\mathrm{T}}\mathbf{A}\mathbf{s}_0}{\mathbf{s}_0^{\mathrm{T}}\mathbf{A}\mathbf{s}_0} = -\frac{-8}{80} = \frac{1}{10}. \tag{20.28}$$

Then the search direction is

$$\mathbf{s}_1 = \mathbf{r}_1 + \beta_0\mathbf{s}_0 = \tfrac{1}{5}\{-1, \quad 10, \quad 3\}^{\mathrm{T}}. \tag{20.29}$$

The constant α_1 is

$$\alpha_1 = \frac{\mathbf{r}_1^T \mathbf{r}_1}{\mathbf{s}_1^T \mathbf{A} \mathbf{s}_1} = \frac{4}{36/5} = \frac{5}{9}. \tag{20.30}$$

Therefore, our next guess is

$$\mathbf{x}_2 = \mathbf{x}_1 + \alpha_1 \mathbf{s}_1 = \tfrac{1}{9}\{-1, \quad 19, \quad 39\}^T. \tag{20.31}$$

The subsequent calculations go as follows:

$$\begin{aligned}
\mathbf{r}_2 &= \mathbf{b} - \mathbf{A}\mathbf{x}_2 = \tfrac{4}{9}\{3 \quad 0 \quad 1\}^T, \\
\beta_1 &= \frac{\mathbf{r}_2^T \mathbf{A} \mathbf{s}_1}{\mathbf{s}_1^T \mathbf{A} \mathbf{s}_1} = \frac{40}{81}, \\
\mathbf{s}_2 &= \mathbf{r}_2 + \beta_1 \mathbf{s}_1 = \tfrac{20}{81}\{5 \quad 4 \quad 3\}^T, \\
\alpha_2 &= \frac{\mathbf{r}_2^T \mathbf{r}_2}{\mathbf{s}_1^T \mathbf{A} \mathbf{s}_1} = \frac{9}{10}, \\
\mathbf{x}_3 &= \{1 \quad 3 \quad 5\}^T, \\
\mathbf{r}_3 &= \mathbf{b} - \mathbf{A}\mathbf{x}_3 = 0.
\end{aligned} \tag{20.32}$$

Consequently, the solution $\mathbf{x}_3$ is the final solution which also happens to be the exact solution. Notice that it took three iterations to reach the solution since the matrix **A** is a three-dimensional matrix.

We rarely use the CG method to solve small systems of equations. This method really shines when we deal with large and narrowly banded systems of equations.

The Method of Lagrange Multiplier and Newton's Method

Consider the following constrained problem:

$$\begin{aligned}
&\text{minimize} \quad f(x,y,z) = x^2 + 2y^2 + 3z^2, \\
&\text{subject to} \quad x^2 + y^2 + z^2 = 1, \\
&\qquad\qquad\quad x + y + z = 0.
\end{aligned} \tag{20.33}$$

We can solve this problem by constructing the Lagrange function

$$L(x,y,z,\lambda_1,\lambda_2) = x^2 + 2y^2 + 3z^2 + \lambda_1(x^2 + y^2 + z^2 - 1) + \lambda_2(x + y + z) \tag{20.34}$$

and by finding the root vector $\mathbf{x} = \{\bar{x}, \bar{y}, \bar{z}, \bar{\lambda}_1, \bar{\lambda}_2\}^{\mathrm{T}}$ that cancels out the components of the gradient of L. In other words, we want to find $\mathbf{x}$ that satisfies

$$\left\{\begin{array}{r} 2x + 2\lambda_1 x + 1 = 0 \\ 4y + 2\lambda_1 y + 1 = 0 \\ 6z + 2z + 1 = 0 \\ x^2 + y^2 + z^2 - 1 = 0 \\ x + y + z = 0 \end{array}\right\}. \tag{20.35}$$

Instead of solving (20.35) by hand, we are going to approximate $\mathbf{x}$ numerically by means of Newton's method. In other words, we are going to guess a solution and refine it successively. As we have seen in Chapter 19, Newton's method for a 1-D function $f(x)$ is given by

$$x_{k+1} = x_k - \frac{f'(x_k)}{f''(x_k)} \tag{20.36}$$

which finds the roots of $f'(x) = 0$. For higher dimensional functions, the corresponding form of (20.36) is

$$\mathbf{x}_{k+1} = \mathbf{x}_k - \mathbf{H}_k^{-1} \nabla L_k \tag{20.37}$$

where the matrix $\mathbf{H}$ is known as the **Hessian** whose components h_{ij} are defined by

$$h_{ij} = \frac{\partial^2 L(\mathbf{x})}{\partial x_i \partial x_j}, \quad i,j = 1,2,\ldots N \tag{20.38}$$

where N is the dimensions of the function L. The Hessian for the current problem is

$$\mathbf{H}=\begin{bmatrix} 2(1+\bar{\lambda}_1) & 0 & 0 & 2\bar{x}_1 & 1 \\ 0 & 4(2+\bar{\lambda}_1) & 0 & 2\bar{x}_2 & 1 \\ 0 & 0 & 2(\bar{\lambda}_1+9x_3) & 2\bar{x}_3 & 1 \\ 2\bar{x}_1 & 2\bar{x}_2 & 2\bar{x}_3 & 0 & 0 \\ 1 & 1 & 1 & 0 & 0 \end{bmatrix}. \quad (20.39)$$

Notice the symmetry in **H**. We may take advantage of symmetry to reduce the computation time, but we do not worry here. ∇L is the gradient of L whose components are

$$\nabla L=\left\{\frac{\partial L}{\partial x_1},\ \frac{\partial L}{\partial x_2},\ \cdots,\ \frac{\partial L}{\partial x_N}\right\}^{\mathrm{T}}. \quad (20.40)$$

To evaluate (20.37) directly is a very expensive operation because of the presence of the inverse of **H** which needs to be updated at each iteration. Therefore, we are going to rearrange (20.37) as

$$\begin{aligned} \mathbf{H}_k\Delta\mathbf{x}_k &= -\nabla L_k, \\ \Delta\mathbf{x}_k &= \mathbf{x}_{k+1}-\mathbf{x}_k. \end{aligned} \quad (20.41)$$

The advantage of using (20.41) instead of (20.37) is that now it is a system of simultaneous linear equations. Solving (20.41) is much cheaper and faster.

Let us compute the first iteration. We choose the starting point at $\mathbf{x}_0=\{1 \quad 2 \quad 1 \quad 1 \quad 1\}^{\mathrm{T}}$. After substituting $\mathbf{x}_0$ into (20.35) and (20.39), we have the first system of equations

$$\begin{bmatrix} 4 & 0 & 0 & 2 & 1 \\ 0 & 6 & 0 & 4 & 1 \\ 0 & 0 & 20 & 2 & 1 \\ 2 & 4 & 2 & 0 & 0 \\ 1 & 1 & 1 & 0 & 0 \end{bmatrix}\Delta\mathbf{x}_0=-\begin{Bmatrix} 5 \\ 13 \\ 12 \\ 5 \\ 4 \end{Bmatrix}. \quad (20.42)$$

The solution is

$$\Delta\mathbf{x}_0=\tfrac{1}{24}\{-103,\ \ 36,\ \ -29,\ \ -410,\ \ 1112\}^{\mathrm{T}}. \quad (20.43)$$

Therefore, the next solution is

$$\mathbf{x}_1 = \mathbf{x}_0 + \Delta\mathbf{x} = \tfrac{1}{24}\{-79, \quad 84, \quad -5, \quad -386, \quad 1136\}^{\mathrm{T}}, \tag{20.44}$$

and so on. This procedure is repeated until the norm of the gradient ∇L becomes sufficiently small.

Tables 20.1–20.4 show the results starting at different points. We notice that they converge to different points, and there are two sets of roots for each point. We can try several more starting points and find that they converge to either point and either root. We conclude, therefore, that the minimum value is about 1.42265 and the roots are

$$\{x, \quad y, \quad z\} = \left\{\begin{matrix} -0.78868, & 0.57735, & 0.21132 \\ 0.78868, & -0.57735, & -0.21132 \end{matrix}\right\}. \tag{20.45}$$

(20.45) agrees well with the analytical solution

$$\begin{aligned} &\text{minimum value } = 2 - \frac{\sqrt{3}}{3} \cong 1.4226 \\ &\{x, \quad y, \quad z\} = \frac{\sqrt{2+\sqrt{3}}}{\sqrt{6}} \left\{\begin{matrix} -1, & -\left(1-\sqrt{3}\right), & -\left(-2+\sqrt{3}\right) \\ 1, & \left(1-\sqrt{3}\right), & \left(-2+\sqrt{3}\right) \end{matrix}\right\}. \end{aligned} \tag{20.46}$$

We also find during the process the maximum value and the corresponding roots as

$$\begin{aligned} &\text{maximum value } = 2.5774 \\ &\{x, \quad y, \quad z\} = \left\{\begin{matrix} -0.21132, & -0.57735, & 0.78868 \\ 0.21132, & 0.57735, & -0.78868 \end{matrix}\right\} \end{aligned} \tag{20.47}$$

Table 20.1. Newton's method result #1.

k	x	y	z	λ1	λ2	f	$\lvert\nabla L\rvert$
0	-8.0000	9.0000	3.0000	-1.0000	-2.0000	253.0000	
1	-4.9893	3.7697	1.2196	-1.2217	-3.5479	57.7760	6.4835E+00
2	-2.5723	1.8919	0.6803	-1.3247	-2.1684	15.1640	3.4018E+00
3	-1.4081	1.0312	0.3769	-1.3783	-1.1900	4.5357	1.7743E+00
4	-0.9250	0.6771	0.2478	-1.4074	-0.7819	1.9568	7.3682E-01
5	-0.7987	0.5847	0.2140	-1.4206	-0.6752	1.4591	1.9282E-01
6	-0.7887	0.5774	0.2113	-1.4226	-0.6667	1.4229	1.5341E-02
7	-0.7887	0.5774	0.2113	-1.4226	-0.6667	1.4226	9.9595E-05
8	-0.7887	0.5774	0.2113	-1.4226	-0.6667	1.4226	4.3711E-09
9	-0.7887	0.5774	0.2113	-1.4226	-0.6667	1.4226	7.2456E-17

Table 20.2. Newton's method result #2.

k	x	y	z	λ1	λ2	f	$\lvert\nabla L\rvert$
0	2.0000	1.0000	1.0000	10.0000	20.0000	9.0000	
1	3.5000	-1.8200	-1.6800	-50.3400	164.3600	27.3420	1.5652E+02
2	1.8454	-0.9508	-0.8947	-24.5860	1.8267	7.6148	1.6457E+02
3	1.1035	-0.5568	-0.5467	-10.7800	1.1002	2.7345	1.3855E+01
4	0.8540	-0.4056	-0.4484	-3.6051	0.8683	1.6614	7.1854E+00
5	0.8103	-0.1848	-0.6255	-1.6936	0.9572	1.8987	1.9348E+00
6	1.0116	-0.9336	-0.0780	-1.2759	0.7264	2.7847	1.0624E+00
7	0.8368	-0.6297	-0.2070	-1.3916	0.6959	1.6218	3.9227E-01
8	0.7904	-0.5791	-0.2114	-1.4208	0.6679	1.4294	7.9782E-02
9	0.7887	-0.5774	-0.2113	-1.4226	0.6667	1.4227	3.2948E-03
10	0.7887	-0.5774	-0.2113	-1.4226	0.6667	1.4226	5.5765E-06
11	0.7887	-0.5774	-0.2113	-1.4226	0.6667	1.4226	1.4429E-11
12	0.7887	-0.5774	-0.2113	-1.4226	0.6667	1.4226	6.2182E-17

Table 20.3. Newton's method result #3.

k	x	y	z	λ1	λ2	f	$\lvert\nabla L\rvert$
0	-1.0000	-2.0000	-1.0000	-1.0000	-1.0000	12.0000	
1	3.5000	-3.5000	0.0000	-4.5000	-7.0000	36.7500	8.4705E+00
2	1.6696	-1.9732	0.3036	-2.9605	0.9107	10.8510	8.4097E+00
3	0.3972	-1.4688	1.0716	-2.3595	-0.4496	7.9177	2.1622E+00
4	-0.4784	-0.7515	1.2299	-2.5632	-1.1389	5.8962	1.3501E+00
5	-0.2307	-0.6354	0.8661	-2.5735	-0.7312	3.1112	6.1106E-01
6	-0.2123	-0.5799	0.7922	-2.5770	-0.6696	2.6003	1.1268E-01
7	-0.2113	-0.5774	0.7887	-2.5773	-0.6667	2.5774	5.3297E-03
8	-0.2113	-0.5774	0.7887	-2.5774	-0.6667	2.5774	1.1852E-05
9	-0.2113	-0.5774	0.7887	-2.5774	-0.6667	2.5774	5.9281E-11
10	-0.2113	-0.5774	0.7887	-2.5774	-0.6667	2.5774	6.9850E-17

Table 20.4. Newton's method result #4.

k	x	y	z	λ1	λ2	f	$\lvert\nabla L\rvert$
0	1.0000	2.0000	1.0000	1.0000	1.0000	12.0000	
1	-2.3333	3.5000	-1.1667	-14.1670	39.6670	34.0280	4.1752E+01
2	-1.1643	1.8577	-0.6934	-7.6640	-0.3139	9.7005	4.0559E+01
3	-0.6042	1.1208	-0.5167	-4.2311	-0.0583	3.6781	3.5690E+00
4	-0.2713	0.8674	-0.5961	-2.6010	0.2166	2.6444	1.7071E+00
5	0.2381	0.6924	-0.9305	-2.5703	0.7791	3.6129	8.4806E-01
6	0.2144	0.5856	-0.8000	-2.5763	0.6763	2.6520	1.9892E-01
7	0.2114	0.5774	-0.7888	-2.5773	0.6667	2.5779	1.7224E-02
8	0.2113	0.5774	-0.7887	-2.5774	0.6667	2.5774	1.2337E-04
9	0.2113	0.5774	-0.7887	-2.5774	0.6667	2.5774	6.4155E-09
10	0.2113	0.5774	-0.7887	-2.5774	0.6667	2.5774	1.3210E-16

which also agrees with the analytical solution

$$\begin{aligned} &\text{maximum value } = 2+\frac{\sqrt{3}}{3} \cong 2.5774 \\ &\{x, \quad y, \quad z\} = \frac{\sqrt{2-\sqrt{3}}}{\sqrt{6}} \begin{Bmatrix} -1, & -(1+\sqrt{3}), & (2+\sqrt{3}) \\ 1, & (1+\sqrt{3}), & -(2+\sqrt{3}) \end{Bmatrix}. \end{aligned} \tag{20.48}$$

We ask ourselves, "How do we know that the solution by Newton's method is the solution we are after?" Usually, Newton's method, even if it works, converges to a local extreme point. We cannot say the converged solution is the local minimum or maximum. In many cases, however, we do have a good idea where the minimum point is so that all we have to do is search in that region.

As the size of the problem increases, the manual computation becomes progressively difficult. Thus numerical methods such as Newton's method become essential.

Solution of Nonlinear Equations by Newton's Method

We can extend the idea of applying Newton's method to constrained problems to the idea of applying it to systems of nonlinear equations. Consider the following problem

$$\begin{Bmatrix} f_1(x_1, x_2, \ldots, x_N) = g_1 \\ f_2(x_1, x_2, \ldots, x_N) = g_2 \\ \vdots \\ f_N(x_1, x_2, \ldots, x_N) = g_N \end{Bmatrix} \tag{20.49}$$

where f_i are nonlinear functions of N arguments. We can rewrite (20.49) as

$$\begin{Bmatrix} f_1(x_1, x_2, \ldots, x_N) - g_1 = 0 \\ f_2(x_1, x_2, \ldots, x_N) - g_2 = 0 \\ \vdots \\ f_N(x_1, x_2, \ldots, x_N) - g_N = 0 \end{Bmatrix}. \tag{20.50}$$

We notice immediately that (20.50) has the same form as (20.35) which means we can apply Newton's method (20.41) to solve. In other words,

$$\begin{aligned} \mathbf{H}_k \Delta \mathbf{x}_k &= -\mathbf{f}_k \\ \Delta \mathbf{x}_k &= \mathbf{x}_{k+1} - \mathbf{x}_k \end{aligned} \tag{20.51}$$

where

$$\begin{aligned} \mathbf{H}\colon\ h_{ij} &= \frac{\partial^2 f}{\partial x_i \partial x_j}, \quad i, j = 1, 2, \ldots, N, \\ \mathbf{f} &= \{f_1 - \mathbf{g}_1 \quad f_2 - \mathbf{g}_2 \quad \cdots \quad f_N - \mathbf{g}_N\}. \end{aligned} \tag{20.52}$$

The procedure is exactly the same the one for the constrained problem in the previous section.

Solution of Integral Equations
Example 20.3

We have seen Ritz's and Galerkin's methods for solving differential equations. In this section, we are going to apply these techniques to solve integral equations.

Consider the following problem:

$$\begin{aligned} \int_0^1 xyu(y)\,dy + \tfrac{1}{2}u(x) &= 2\sin(x), \quad 0 \le x \le 1, \\ u(0) &= 0, \\ u(1) &= \tfrac{4}{5}[6\cos(1) - \sin(1)]. \end{aligned} \tag{20.53}$$

First, we apply Galerkin's method. Let us assume the trial function

$$u(x) \cong \bar{u}(x) = \frac{4[6\cos(1) - \sin(1)]}{5}x + c_1 x(x-1) \tag{20.54}$$

where c_1 is an undetermined constant. The weight function for (20.54) is $x(x-1)$. We now apply Galerkin's method

$$\int_0^1 \left[\int_0^1 xy\bar{u}(y)\,dy + \tfrac{1}{2}\bar{u}(x) - 2\sin(x)\right] x(x-1)\,dx = 0. \tag{20.55}$$

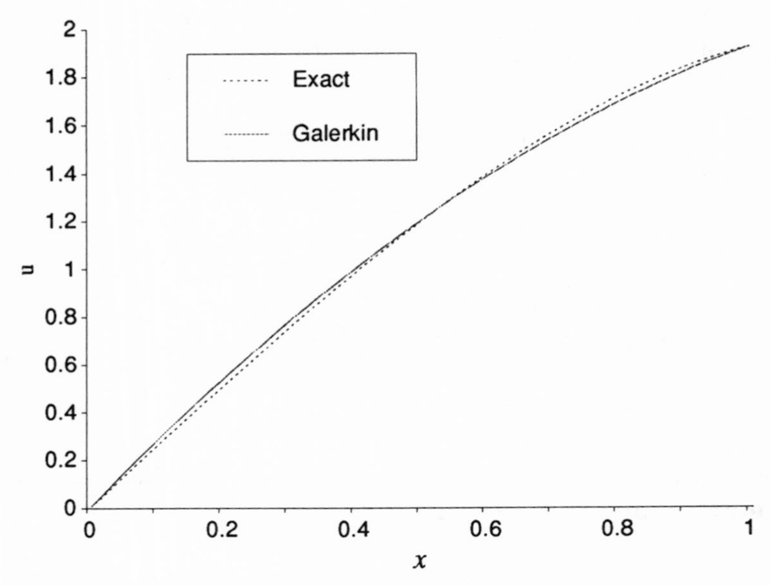

Figure 20.1. Galerkin's solution and the exact solution.

After solving for c_1 from (20.55), the solution is approximately

$$\overline{u}(x) = -0.95279x^2 + 2.87306x, \tag{20.56}$$

which agrees well with the exact solution

$$u(x) = \frac{24x[\cos(1) - \sin(1)]}{5} - 4\sin(x) \tag{20.57}$$

as shown in Fig. 20.1.

Next, we try Ritz's method. The least-squares method described in Chapter 3 can be applied here. The substitution of the same trial function (20.54) into (20.53) yields an error

$$\int_0^1 xyu(y)\,dy + \tfrac{1}{2}u(x) - 2\sin(x) = \varepsilon(x). \tag{20.58}$$

We are going to minimize the squared magnitude of ε over the domain. In other words, we are going to extremize

$$\int_0^1 [\varepsilon(x)]^2\,dx = \int_0^1 \left[\int_0^1 xyu(y)\,dy + \tfrac{1}{2}u(x) - 2\sin(x)\right]^2 dx \tag{20.59}$$

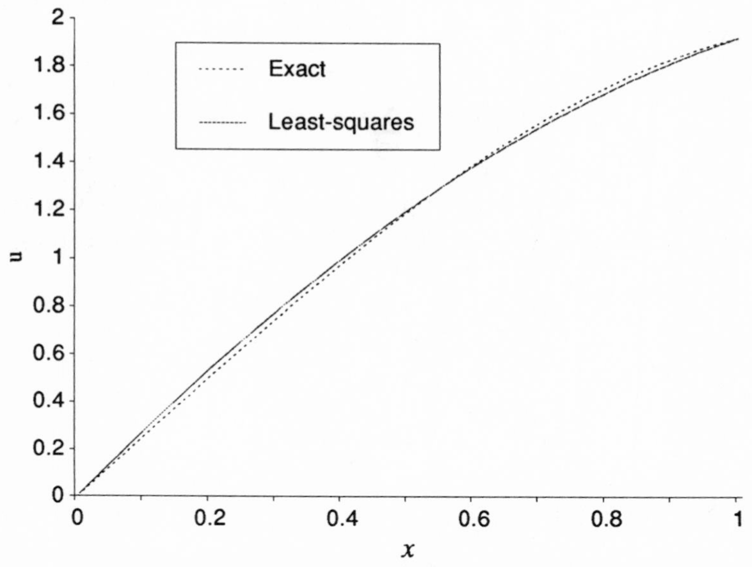

Figure 20.2. Least-squares' solution and the exact solution.

with respect to c_1 of (20.54). This yields

$$\bar{u}(x) = -0.965872x^2 + 2.88615x \tag{20.60}$$

which closely agrees with the exact solution as shown in Fig. 20.2.

Next, we take a look at a slightly odd problem.

Example 20.4

Consider

$$\int_a^b k(x, x')u(x)\,ds(x) = b(x'), \quad c \le x' \le d, \tag{20.61}$$

where a and c and b and d may or may not be equal. This problem, unlike the previous one, x and x', might not share the same domain. Furthermore, (20.61) maybe difficult to integrate because of the nature of the integrand or the geometry of the domain. Such a problem is ill-posed, and we cannot usually expect a good solution by the standard techniques. Because we cannot expect a good solution, we attempt to obtain at least a *meaningful* solution by means of the variational principles.

First, we approximate (20.61) numerically by a Nth-order quadrature

$$\int_a^b k(x,x')u(x)ds(x) \cong \sum_{k=1}^{N} k(x_k,x')u(x_k)w_k \tag{20.62}$$

where x_k are sample points and w_k are weights. Since we are now solving for N unknowns $u(x_k)$, let us choose N arbitrary x' points

$$\left\{\begin{array}{c} \sum_{k=1}^{N} k(x_k,x_1')u(x_k)w_k = b(x_1') \\ \sum_{k=1}^{N} k(x_k,x_2')u(x_k)w_k = b(x_2') \\ \vdots \\ \sum_{k=1}^{N} k(x_k,x_N')u(x_k)w_k = b(x_N') \end{array}\right\}. \tag{20.63}$$

Thus we have a system of simultaneous equations, and by solving (20.63), we can compute u at the sample points.

Unfortunately, this procedure works for well-posed problems such as Example 20.3 only. When the problem is ill-posed, it is very rare that (20.63) can be solved directly because it is usually highly singular. Another difficulty may be that in some situations we cannot choose exactly N number of points. Sometimes we have too many, and sometimes we have too few so that the number of equations may not be necessarily the number of unknowns.

In order to overcome these difficulties, we execute the following operations. First, we write (20.63) as

$$\mathbf{Ku} = \mathbf{b} \tag{20.64}$$

where

$$\mathbf{K} = \begin{bmatrix} k(x_1,x_1')w_1 & k(x_2,x_1')w_2 & \cdots & k(x_N,x_1')w_N \\ k(x_1,x_2')w_1 & k(x_2,x_2')w_2 & \cdots & k(x_N,x_2')w_N \\ \vdots & \vdots & \ddots & \vdots \\ k(x_1,x_M')w_1 & k(x_2,x_M')w_2 & \cdots & k(x_M,x_N')w_N \end{bmatrix},$$

$$\mathbf{u} = \{u(x_1) \quad u(x_2) \quad \cdots \quad u(x_N)\}^{\mathrm{T}},$$

$$\mathbf{b} = \{b(x_1) \quad b(x_2) \quad \cdots \quad u(x_M)\} \tag{20.65}$$

If the rank of **K** is N, then we can apply the least-squares method introduced at the beginning of this chapter

$$\mathbf{u} = \left(\mathbf{K}^{\mathrm{T}}\mathbf{K}\right)^{-1}\mathbf{K}^{\mathrm{T}}\mathbf{b}. \tag{20.66}$$

Then we have a solution in the least-squares sense.

What if the rank of **K** is less than N? In this case, we rephrase the problem (20.64) as

$$\begin{aligned} &\text{minimize} \quad \mathbf{u}^{\mathrm{T}}\mathbf{u}, \\ &\text{subject to} \quad (\mathbf{Ku}-\mathbf{b})^{\mathrm{T}}(\mathbf{Ku}-\mathbf{b}) < \varepsilon. \end{aligned} \tag{20.67}$$

In other words, we want the smallest solution in squared magnitude while at the same time keeping the squared magnitude of the approximation error at less than some prescribed tolerance ε. The Lagrangian function for this problem is

$$L(\mathbf{u},\lambda) = \mathbf{u}^{\mathrm{T}}\mathbf{u} + \lambda\left[(\mathbf{Ku}-\mathbf{b})^{\mathrm{T}}(\mathbf{Ku}-\mathbf{b}) - \varepsilon\right]. \tag{20.68}$$

When (20.68) is extremized with respect to **u**, we have

$$\left(\mathbf{K}^{\mathrm{T}}\mathbf{K} + \alpha\mathbf{I}\right)\mathbf{u} = \mathbf{K}^{\mathrm{T}}\mathbf{b} \tag{20.69}$$

where α is the reciprocal of λ, and **I** is the identity matrix. We notice that when α is zero, (20.69) becomes the normal equation (20.5). The reason we cannot apply the least-squares method to the current problem is that $\mathbf{K}^{\mathrm{T}}\mathbf{K}$ is a singular matrix. However, when $\alpha\mathbf{I}$, with a sufficient value of α, is added to $\mathbf{K}^{\mathrm{T}}\mathbf{K}$, $\mathbf{K}^{\mathrm{T}}\mathbf{K}+\alpha\mathbf{I}$ becomes no longer singular; therefore, we can solve (20.69). This method is called the **regularization method.** Let us take a look at a specific example.

Example 20.5

Consider the following problem

$$\begin{bmatrix} 1 & 1 \\ 2 & 2 \\ 3 & 3 \end{bmatrix} \begin{Bmatrix} x_1 \\ x_2 \end{Bmatrix} = \begin{Bmatrix} -1 \\ -2 \\ -3 \end{Bmatrix}. \tag{20.70}$$

This problem has no unique solution because any solution

$$\begin{Bmatrix} x_1 \\ x_2 \end{Bmatrix} = -\frac{1}{2}\begin{Bmatrix} 1 \\ 1 \end{Bmatrix} + k\begin{Bmatrix} 1 \\ -1 \end{Bmatrix} \tag{20.71}$$

where k is an arbitrary constant can satisfy (20.70). Furthermore, because the rank of the coefficient matrix is 1, we cannot use the least-squares method.

Let us apply (20.69) for this problem. We have

$$\begin{gathered} \left\{ \begin{bmatrix} 1 & 2 & 3 \\ 1 & 2 & 3 \end{bmatrix} \begin{bmatrix} 1 & 1 \\ 2 & 2 \\ 3 & 3 \end{bmatrix} + \alpha \mathbf{I} \right\} \begin{Bmatrix} x_1 \\ x_2 \end{Bmatrix} = \begin{bmatrix} 1 & 2 & 3 \\ 1 & 2 & 3 \end{bmatrix} \begin{Bmatrix} -1 \\ -2 \\ -3 \end{Bmatrix} \\ \begin{bmatrix} 1+\alpha & 1 \\ 1 & 1+\alpha \end{bmatrix} \begin{Bmatrix} x_1 \\ x_2 \end{Bmatrix} = \begin{Bmatrix} 1 \\ 1 \end{Bmatrix} \\ \begin{Bmatrix} x_1 \\ x_2 \end{Bmatrix} = -\frac{1}{2+\alpha} \begin{Bmatrix} 1 \\ 1 \end{Bmatrix}. \end{gathered} \tag{20.72}$$

As $\alpha \to 0$, we have the particular solution for the problem. The regularization method successfully finds the meaningful part of the nonunique solution. Let us take a look at a more complicated problem.

Example 20.6

An elastic body is loaded as shown in Fig. 20.3. Assuming a static equilibrium, we have the following boundary conditions:

$$\begin{aligned} u(x) &= \begin{cases} \bar{u}_2, & x \in \Gamma_2, \\ \bar{u}_6, & x \in \Gamma_6, \end{cases} \\ t(x) &= \begin{cases} \bar{t}_2, & x \in \Gamma_2, \\ \bar{t}_6, & x \in \Gamma_6, \end{cases} \end{aligned} \tag{20.73}$$

where u and t are displacement and traction, respectively. We would like to compute the displacements u_1 and u_2 along Γ_1. We omit the derivation details here. The integral equation we are going to solve is

$$\int_\Gamma C_{ijkl} G_{km,l}(x,x') u_i(x) n_j \, ds(x) + \alpha u_m(x') - \int_\Gamma G_{im}(x,x') t_i(x) \, ds(x) = b_m(x'), \tag{20.74}$$

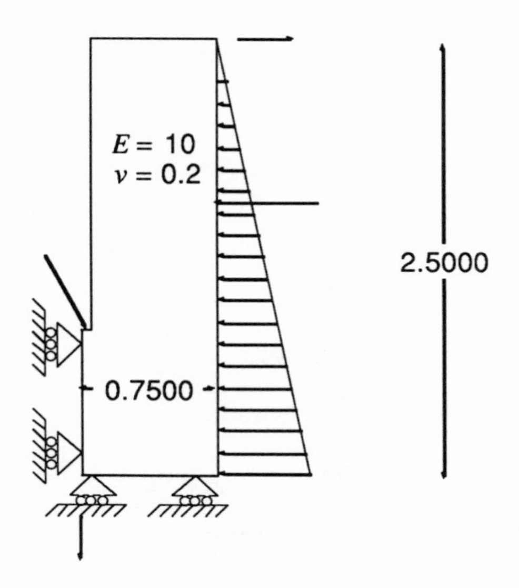

Figure 20.3. Example 20.6.

where C_{ijkl} are the material constants, G_{ij} is the Green's tensor, n_i is the outer normal vector, and

$$\Gamma = \Gamma_1 + \Gamma_3 + \Gamma_4 + \Gamma_5$$

$$b_m(x') = -\int_{\Gamma_2+\Gamma_6} C_{ijkl} G_{km,l}(x,x') u_i(x) n_j \, ds(x) - \alpha u_m(x') + \int_{\Gamma_2+\Gamma_6} G_{im}(x,x') t_i(x) \, ds(x),$$

$$\alpha = \begin{cases} \frac{1}{2}, & x' \in \text{smooth boundary}, \\ \frac{1}{4}, & x' \in \text{corner}. \end{cases}$$

For more details of this equation, consult a book on the boundary element method such as the one by C. A. Brebbia, J. C. F. Tells, and L. C. Wrobel (*Boundary Element Techniques,* Springer-Verlag, New York, 1984).

This problem is not a standard boundary-value problem because the standard ones require either the displacement or the traction to be prescribed to the entire boundary. This problem is clearly ill-posed.

We can convert (20.74) into the form of (20.64) by subdividing each boundary into smaller ones and apply the regularization method. Because we have very few knowns, we cannot expect a very good result, however.

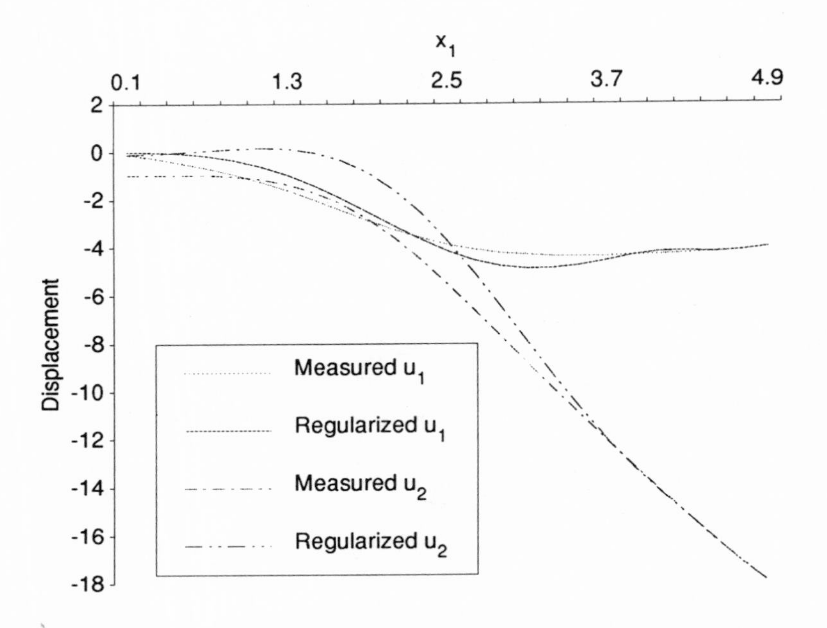

Figure 20.4. The regularized solutions and the measured solutions.

Because the calculations are lengthy and complicated, we omit the details. Figure 20.4 shows the regularized solutions and the measured solutions. We note that the errors are still significantly large, but we manage to obtain the approximate behavior of the measured solution.

21

Bounds for the Overall Properties of Anisotropic Composites

Basic Theory

This chapter gives an account of the application of variational methods in the theory of composite materials. The theory is best developed by exploiting some basic concepts from convex analysis as set out, for instance, by J. Van Tiel (*Convex Analysis*, Wiley, New York, 1984). The elementary facts that are needed are recorded in this chapter but the reader should refer to Van Tiel for derivations. The level of exposition remains simple and informal but the ideas that are represented may perhaps give the reader a first impression of the more sophisticated theory developed, for example, by I. Ekeland and R. Temam (*Convex Analysis and Variational Problems,* North-Holland, Amsterdam, 1976).

The basic problem to be addressed is that of estimating the mean energy density, $\tilde{W}(\bar{e})$, in a sample of material, when it is subjected to loads that generate within it a strain field whose mean value is $\bar{e}$. If the dimensions of the sample are large in comparison with typical microstructural dimensions, then it can be considered to be a *representative volume element* for the composite, and the energy function $\tilde{W}(\bar{e})$ can be taken to define its stress–strain response in any situation in which any length scale associated with the loading is greater than the scale of the representative volume element. The stress and strain, of course, do not vary so smoothly, if viewed on a smaller scale, on account of the fine-scale heterogeneity of the composite, and if such small-scale variations are of interest then more information than that delivered by $\tilde{W}(\bar{e})$ will be required. The same situation applies, however, to the whole of continuum mechanics: No continuum theory can apply down to arbitrarily small-length scales.

The problem of estimating $\tilde{W}(\bar{e})$ is now addressed, without further ado. The sample of composite will be taken to occupy a domain Ω and, for convenience, units will be chosen so that Ω has unit volume. The composite is assumed to be made from n constituents, firmly bonded together across interfaces (imperfect bonding can be important but this exposition is restricted to the simpler case of perfect bonding, for which the theory is more highly developed). Material of type r is taken to have energy density function $W_r(e)$. The energy density function for the composite depends on position x, because different materials are present at different points. It takes the form

$$W(e,x) = \sum_{r=1}^{n} W_r(e) f_r(x), \tag{21.1}$$

where $f_r(x)$ is the characteristic function of the region occupied by material of type r, taking the value 1 if x is in region r and 0 otherwise, so that the sum of the function f_r equals 1, for any x. The strain e has components e_{ij}, as defined in Chapter 11. Only infinitesimal strains are considered but there is no need to assume linear behavior, so the functions $W_r(e)$ do not (yet) have to be quadratic. They will, however, be assumed to be convex, and so to have the properties

$$W_r[\lambda e_1 + (1-\lambda)e_2] \geq \lambda W_r(e_1) + (1-\lambda) W_r(e_2), \tag{21.2}$$

for all $\lambda \in (0,1)$, and

$$W_r(e_2) \geq W_r(e_1) + (e_2 - e_1) \cdot W_r'(e_1), \tag{21.3}$$

both for all e_1 and e_2. The inequality (21.2) provides a basic definition of convexity; (21.3) is equivalent, so long as W_r is differentiable. In (21.3), a symbolic notation has been introduced for conciseness: W'_r signifies a "vector" with components $\partial W_r / \partial e_{ij}$ and the dot signifies a "scalar product," the sum extending over the values 1,2,3 of the suffixes i and j, which are suppressed in (21.3).

The mean strain $\bar{e}$ is generated in the sample by imposing the boundary condition

$$u_i = \bar{e}_{ij} x_j, \quad x \in \partial\Omega, \tag{21.4}$$

where u denotes displacement. Equilibrium is attained when the field u minimizes the strain energy, over displacements that conform to (21.4). Thus, $\tilde{W}(\bar{e})$ is defined by

$$\tilde{W}(\bar{e}) = \inf_e \int_\Omega W(e,x)\, dx, \tag{21.5}$$

where the infimum is taken over strain fields derived from displacements that satisfy (21.4). An elementary bound for $\tilde{W}$ follows immediately from its definition, by choosing u to have the form (21.4) *throughout* Ω:

$$\tilde{W}(\bar{e}) \leq \int_{\Omega} W(\bar{e},x)dx = \overline{W}(\bar{e}). \tag{21.6}$$

An alternative expression for $\overline{W}$ is

$$\overline{W}(\bar{e}) = \sum_{r=1}^{n} c_r W_r(\bar{e}).$$

Hence,

$$c_r = \int_{\Omega} f_r(x)dx$$

denotes the volume fraction occupied by material of type r.

An elementary lower bound can also be obtained, by use of the complementary energy principle. This is now developed in the form needed; an extension of the argument will subsequently lead to better bounds. First, the complementary energy density, dual to $W(e,x)$, is defined as

$$W^*(\sigma,x) = \sup_{e}[\sigma \cdot e - W(e,x)]. \tag{21.7}$$

Fenchel's inequality

$$W(e,x) + W^*(\sigma,x) \geq \sigma \cdot e \tag{21.8}$$

follows directly from the definition of W^*. Equality in (21.8) is attained when

$$\sigma = W'(e,x).$$

The definition (21.7) permits the immediate deduction that W^* is convex in σ. If W is convex in e, equality in (21.8) is also attained when

$$e = W^{*\prime}(\sigma, x)$$

and

$$W(e,x) = W^{**}(e,x) = \sup_{\sigma}\left[\sigma \cdot e - W^{*}(\sigma,x)\right]. \tag{21.9}$$

It follows, directly from the definition (21.5) of $\tilde{W}$ and Fenchel's inequality (21.8), that

$$\tilde{W}(\bar{e}) \geq \inf_{e} \int_{\Omega} \left[\sigma \cdot e - W^{*}(\sigma,x)\right] dx, \tag{21.10}$$

for any σ. The infimum is $-\infty$ unless σ has zero divergence, when the single term involving e has the value $\bar{\sigma} \cdot \bar{e}$. Thus, by maximizing the bound (21.10) with respect to σ,

$$\tilde{W}(\bar{e}) \geq \sup_{\bar{\sigma}}\left[\bar{\sigma} \cdot e - \tilde{W}^{*}(\bar{\sigma})\right], \tag{21.11}$$

where

$$\tilde{W}^{*}(\bar{\sigma}) \geq \inf_{\sigma} \int_{\Omega} W^{*}(\sigma,x)\, dx, \tag{21.12}$$

the infimum being taken over fields σ with zero divergence and prescribed mean value $\bar{\sigma}$.

The elementary lower bound follows by choosing $\sigma = \bar{\sigma}$ throughout Ω. This gives

$$\left(\overline{W}^{*}\right)^{*}(\bar{e}) \leq \tilde{W}(\bar{e}) \leq \overline{W}(\bar{e}), \tag{21.13}$$

where

$$\overline{W}^{*}(\bar{\sigma}) = \sum_{r=1}^{n} c_r W_r^{*}(\bar{\sigma}),$$

having also repeated the upper bound (21.6).

In the special case of linear elastic behavior, the stress–strain behavior of material of type r can be written as

$$\sigma_{ij} = L_{ijkl}^{(r)} e_{kl}, \quad e_{ij} = M_{ijkl}^{(r)} \sigma_{kl}$$

or, symbolically,

$$\sigma = L_r e, \quad e = M_r \sigma,$$

so that L_r and M_r are inverses. Correspondingly,

$$W_r(e) = \tfrac{1}{2} e L_r e, \quad W_r^*(\sigma) = \tfrac{1}{2} \sigma M_r \sigma.$$

Then, the bounds (21.13) yield

$$\bar{e}\left(\sum c_r M_r\right)^{-1} \bar{e} \le \bar{e}\tilde{L}\bar{e} \le \bar{e}\left(\sum c_r L_r\right)\bar{e},$$

where $\tilde{L}$ denotes the overall tensor of elastic moduli, associated with $\tilde{W}$. These lower and upper bounds are, respectively, the estimates of Reuss and Voigt; they were shown to provide bounds on the energy by R. Hill ("The elastic behavior of a crystalline aggregate," *Proc. Phys. Soc.* **A65**, 349–354, 1952).

Improved Bounds

The question is now addressed of finding bounds which make an allowance for more geometrical information on the composite than just volume fractions. The difficulty is in the selection of suitable approximating stress and strain fields for substitution into the variational principles (21.6) and (21.12). One way forward is to introduce a "companion" material, occupying Ω, with energy density function $W_0(e)$. Historically, this was developed by first representing the stress in the form

$$\sigma = W'(e,x) = W_0'(e) + \tau, \tag{2.14}$$

where τ is a "stress polarization," and substituting into the equilibrium equation. More recently, however, in the course of generalizing the approach to deal with nonlinear material behavior, a more concise formulation has been obtained, by working directly from the variational principle (21.6). The core of the idea is to define, without direct physical motivation, the function

$$(W - W_0)^*(\tau, x) = \sup_e\left[\tau \cdot e - (W - W_0)(e,x)\right]. \tag{21.15}$$

The function $(W - W_0)$ may not be convex, but $(W - W_0)^*$ is. The supremum is attained when

$$\tau = (W - W_0)'(e, x)$$

and this is consistent with (21.14). It should be noted, however, that this equation may be satisfied by more than one e, if $W - W_0$ is not convex; in that case, the required e is the one that gives the global maximum, as specified by (21.15). Now following the reasoning whereby the complementary principle (21.12) was developed, the energy principle (21.5), coupled with the Fenchel inequality associated with (21.15), yields

$$\tilde{W}(\bar{e}) \geq \inf_{e} \int_{\Omega} \left[\tau \cdot e + W_0(e) - (W - W_0)^*(\tau) \right] dx, \tag{21.16}$$

for any choice of τ. The inequality (21.16) generalizes, to nonlinear material behavior, the celebrated variational principle of Z. Hashin and S. Shtrikman ("On some variational principles in anisotropic and nonhomogeneous elasticity," *J. Mech. Phys. Solids* **10**, 335–342, 1962). It is easier to work with than (21.5) because the evaluation of the infimum over e only requires consideration of the function W_0 rather than W. In fact, the associated Euler–Lagrange equation is

$$\operatorname{div}\{W_0'(e)\} + \operatorname{div} \tau = 0, \quad x \in \Omega, \tag{21.17}$$

which is to be solved with the boundary condition (21.4). This is an elastic boundary-value problem for the comparison material, which can be chosen to be uniform, although this is not a requirement of the general theory. To make further progress, it is necessary to specialize W_0 to be a quadratic function,

$$W_0(e) = \tfrac{1}{2} e L_0 e,$$

corresponding to a linearly elastic comparison medium with tensor of elastic moduli L_0. The Euler–Lagrange equation (21.17) then becomes

$$\operatorname{div}(L_0 e) + \operatorname{div} \tau = 0, \quad x \in \Omega. \tag{21.18}$$

The solution of (21.18) can be given in terms of a Green's tensor G, which satisfies

$$\operatorname{div}(L_0 \nabla G) + I\delta(x - x') = 0, \quad x \in \Omega,$$

with the boundary condition

$$G(x, x') = 0, \quad x \in \partial\Omega.$$

The strain e associated with the solution of (21.18) follows as

$$e = \bar{e} - \Gamma\tau, \tag{21.19}$$

where the linear operator Γ is defined so that

$$(\Gamma\tau)_{ij}(x) = \int_\Omega \Gamma_{ijkl}(x, x')\tau_{kl}(x')\,dx' \tag{21.20}$$

with

$$\Gamma_{ijkl}(x, x') = \left.\frac{\partial^2 G_{ik}}{\partial x_k \partial x'_l}\right|_{(ij),(kl)},$$

the brackets on the suffixes implying symmetrization. The singularity at $x' = x$ has to be interpreted in the sense of generalized functions. Details are given by J. R. Willis ("Bounds and self-consistent estimates for the overall moduli of anisotropic composites," *J. Mech. Phys. Solids* **25,** 182–202, 1977).

The operator Γ is self-adjoint and has the properties

$$\int_\Omega \Gamma_{ijkl}(x, x')\,dx = 0 \quad \text{and} \quad \Gamma L_0 \Gamma = \Gamma.$$

Substitution of (21.19) into the inequality (21.16), using these properties, gives

$$\tilde{W}(\bar{e}) \geq W_0(\bar{e}) + \int_\Omega \left[\tau \cdot \bar{e} - \frac{1}{2}\tau\Gamma\tau - (W - W_0)^*(\tau)\right] dx. \tag{21.21}$$

This is a bound for any τ. The objective is now to select a convenient set of fields τ over which to optimize the bound. Let τ have the piecewise constant form

$$\tau(x) = \sum_{r=1}^{n} \tau_r f_r(x), \tag{21.22}$$

where the tensor τ_r are constants. The bound (21.21) then becomes

$$\tilde{W}(\bar{e}) \geq W_0(\bar{e}) + \sum_{r=1}^{n} \left\{ c_r \left[\tau_r \cdot \bar{e} - \tfrac{1}{2} \sum_{s=1}^{n} A_{rs} \tau_s - (W_r - W_0)^* (\tau_r) \right] \right\}, \tag{21.23}$$

where A_{rs} is defined so that

$$c_r A_{rs} = \int_{\Omega} \left[f_r(x) \int_{\Omega} \Gamma(x, x') f_s(x') dx' \right] dx \quad \text{(no sum on r)}. \tag{21.24}$$

Each A_{rs} is a tensor of fourth order; suffixes *ijkl* have been suppressed.

The bound (21.23) is maximized with respect to the variables τ_r when

$$(W_r - W_0)^{*\prime} (\tau_r) + \sum_{s=1}^{n} A_{rs} \tau_s = \bar{e}, \tag{21.25}$$

assuming that $(W_r - W_0)^*$ is differentiable. This need not always be the case. The more general form of (21.25) is

$$\gamma_r + \sum_{s=1}^{n} A_{rs} \tau_s = \bar{e}, \tag{21.26}$$

where γ_r is a *subgradient* of $(W_r - W_0)^*$ at τ_r:

$$\gamma_r \in \partial (W_r - W_0)^* (\tau_r), \tag{21.27}$$

which implies that

$$\gamma_r \cdot (\tau - \tau_r) \leq (W_r - W_0)^* (\tau) - (W_r - W_0)^* (\tau_r)$$

for all τ. If $(W_r - W_0)^*$ is differentiable, the only γ_r satisfying (21.27) is $(W_r - W_0)^{*\prime}(\tau_r)$, so (21.25) is recovered. It is appropriate to record, at this point, that, when $W_r - W_0$ is not convex,

$$(W_r - W_0)(\gamma_r) \geq (W_r - W_0)^{**}(\gamma_r) = \sup_{\tau}\left[\tau \cdot \gamma_r - (W_r - W_0)^*(\gamma)\right].$$

The supremum is attained when

$$\tau = \tau_r \in \partial(W_r - W_0)^{**}(\gamma_r). \tag{21.28}$$

Relations (21.27) and (21.28) are satisfied simultaneously. The optimal bound (21.23) can now be written

$$\tilde{W}(\bar{e}) \geq W_0(\bar{e}) + \sum_{r=1}^{n}\left[\tfrac{1}{2}(\bar{e} - \gamma_r) + (W_r - W_0)^{**}(\gamma_r)\right], \tag{21.29}$$

having used (21.26), and observing that

$$(W_r - W_0)^{**}(\gamma_r) = \gamma_r \cdot \tau_r - (W_r - W_0)^*(\tau_r)$$

when (21.28) is satisfied. It is usually the case in practice that $(W_r - W_0)^{**}$ is differentiable, even if $(W_r - W_0)^*$ may not be; then, satisfaction of (21.28) implies

$$\tau_r = (W_r - W_0)^{**\prime}(\gamma_r). \tag{21.30}$$

In summary, therefore, an explicit lower bound for $\tilde{W}(\bar{e})$ is given by (21.29), once (21.26) are solved. They are, in general, nonlinear equations, on account of the relation (21.28) or, equivalently, (21.30), between τ_r and γ_r. The bound makes allowance for the geometrical arrangement of the composite, taking two points at a time, through the tensors A_{rs}.

Upper and Lower Bounds

A formula giving an upper bound can be developed similarly, by noting that

$$W(e) \le \tau \cdot e + W_0(e) + (W_0 - W)^*(-\tau)$$

for any e and τ. It follows from (21.5), therefore, that

$$\tilde{W}(\bar{e}) \le \inf_e \int_\Omega \left[\tau \cdot e + W_0(e) + (W_0 - W)^*(-\tau)\right] dx, \tag{21.31}$$

for any τ. This bound may be compared with the lower bound (21.16). It can be optimized, similarly over fields τ of the form (21.22), to give

$$\tilde{W}(\bar{e}) \le W_0(\bar{e}) + \sum_{r=1}^{n} c_r \left[\tfrac{1}{2}(\bar{e} - \gamma_r) \cdot \tau_r - (W_0 - W_r)^{**}(\gamma_r)\right], \tag{21.32}$$

where γ_r and τ_r satisfy (21.26) but now

$$\gamma_r \in \partial (W_0 - W_r)^*(-\tau_r)$$

and

$$\tau_r = -(W_0 - W_r)^{**\prime}(\gamma_r),$$

if $(W_0 - W_r)^{**}$ is differentiable.

Dual formulations starting from the complementary energy principle (21.12) are also possible. It should be noted that, directly from (21.11), a lower bound for $\tilde{W}^*$ induces a corresponding upper bound for $\tilde{W}$ and an upper bound for $\tilde{W}^*$ induces a lower bound for $\tilde{W}$. Details are not presented, though they are contained, for example, in the paper by J. R. Willis ("Variational estimates for the overall behavior of a nonlinear matrix–inclusion composite," in *Micromechanics and Inhomogeneity, The Toshio Mura Anniversary Volume*, edited by G. J. Weng, M. Taya, and H. Abé, pp.581–597, Springer-Verlag, New York, 1990). The only differences are that each W is replaced by a W^*, stress polarizations τ_r are replaced by strain polarizations η_r, the strainlike variables γ_r are replaced by stresslike variables ζ_r, and the operator Γ is replaced by an operator Δ, where

$$\Delta\eta = L_0(\eta - \bar{\eta}) - L_0 \Gamma L_0 \eta.$$

Which formulation is preferable is a matter of convenience only, since it was shown by D. R. S. Talbot and J. R. Willis ("Variational principles for inhomogeneous nonlinear media," *IMA J. Appl. Math.* **39,** 39–54, 1985) that both deliver the same bounds, if the same linear comparison medium is employed for either, and the polarizations are related so that $\tau_r = -L_0\eta_r$.

A potential problem with the bound formula (21.29), not so far mentioned, is addressed now. To illustrate, suppose that, for some r, $W_r(e)$ behaves asymptotically like $\|e\|^{3/2}$ as $\|e\| \to \infty$. Then the definition (21.15) implies that $(W_r - W_0)^* = +\infty$, for *any* convex quadratic W_0! The bound (21.29) remains true, but the bound is $-\infty$. A nontrivial lower bound can thus be obtained only if W_r grows *at least quadratically* with e, when $\|e\|$ is large, for all r. Similarly, the upper bound (21.32) is nontrivial, only if W_r grows *at most quadratically* with e, when $\|e\|$ is large. Upper and lower bounds are both obtainable, only when each W_r behaves quadratically, asymptotically as $\|e\| \to \infty$. It is remarked that $(W_0^* - W_r^*)^*$ is finite if $(W_r - W_0)^*$ is, so that, in line with the equivalence mentioned above, one formulation yields a nontrivial upper (or lower) bound for $\tilde{W}$ if the other does.

A Two-Component Composite

If the composite has only two components—it may be, for example, a matrix containing reinforcing particles—the tensors A_{rs} can all be expressed in terms of a single integral, upon use of the identity

$$f_1(x) + f_2(x) = 1 \quad \text{for all } x.$$

In fact,

$$A_{rs} = P(\delta_{rs} - c_s), \tag{21.33}$$

where P is defined so that

$$c_1c_2P = \int_\Omega \left[f_1(x)\int_\Omega \Gamma(x,x')f_1(x')dx' \right] dx.$$

Equations (21.26) then take the form

$$\gamma_r + P(\tau_r - \bar{\tau}) = \bar{e}, \tag{21.34}$$

where

$$\bar{\tau} = \sum_{r=1}^{2} c_r \tau_r.$$

Further progress is possible only once W_1 and W_2 are specified. First, consider a linear composite, so that

$$W_r(e) = \tfrac{1}{2} e L_r e.$$

It follows that

$$(W_r - W_0)^*(\tau) = \tfrac{1}{2}\tau (L_r - L_0)^{-1} \tau, \tag{21.35}$$

so long as L_0 is chosen so that $(W_r - W_0)$ is convex; if $(W_0 - W_r)$ is convex, then the same algebraic expression yields $-(W_0 - W_r)^*(-\tau)$. The expression (21.35) for $(W_r - W_0)^*$ is differentiable, so

$$\gamma_r = (L_r - L_0)^{-1} \tau_r$$

and (21.34) give

$$\left[(L_r - L_0)^{-1} + P\right]\tau_r - P\bar{\tau} = \bar{e}. \tag{21.36}$$

The bound formula (21.29) reduces to

$$\tilde{W}(\bar{e}) \geq W_0(\bar{e}) + \tfrac{1}{2}\bar{\tau}\cdot\bar{e},$$

since $(W_r - W_0)^{**} = (W_r - W_0)$ and is homogeneous of degree 2. Solving (21.36) for $\bar{\tau}$ leads finally to the bound

$$\tilde{W}(\bar{e}) \geq \tfrac{1}{2}\bar{e} L_B \bar{e}, \tag{21.37}$$

where

$$L_B = \left\{ \sum c_r \left[I + (L_r - L_0)P \right]^{-1} \right\}^{-1} \sum c_s \left[I + (L_s - L_0)P \right]^{-1} L_s. \qquad (21.38)$$

Expression (21.38) gives a lower bound, for any L_0 for which the quadratic form $\bar{e}(L_r - L_0)\bar{e}$ is positive definite for each r. The upper bound formula (21.32) requires the same algebra; (21.38) correspondingly gives an upper bound, for ant L_0 for which $\bar{e}(L_0 - L_r)\bar{e}$ is positive definite for each r.

The bound formula is a rather general one, in that the individual components of the composite may be anisotropic (though if they are, all material of one type must have the same orientation). Also, no particular symmetry was assumed for the geometry of the composite: Even if L_1 and L_2 are isotropic, the overall behavior need not be. The bound formula reflects this, through the tensor P. The elementary Reuss and Voigt bounds, given earlier, contain only information on volume fractions and cannot reflect geometrical anisotropy.

It is, perhaps, worthy of mention that, when the *geometry* of the composite is isotropic, the tensors A_{rs} have the form (21.33), even when the composite has any number of constituents. The bound formula (21.38) then applies, with the summations extending from 1 to n. In this case too, the tensor P can be given explicitly. The bound is then, essentially, the bound given originally by Z. Hashin and S. Shtrikman ("A variational approach to the theory of the elastic behaviour of multiphase materials," *J. Mech. Phys. Solids* **11,** 127–140, 1963). The form presented here was given first by L. J. Walpole ("On bounds of the overall elastic moduli of inhomogeneous systems," *J. Mech. Phys. Solids* **14,** 151–162, 1966). Bounds (of Hashin–Shtrikman type) for composites with anisotropic geometry were first given by J. R. Willis ("Bounds and self-consistent estimates for the overall moduli of anisotropic composites," *J. Mech. Phys. Solids* **25,** 185–202, 1977). The general two-component bound (21.37) was first given by J. R. Willis ("Elasticity theory of composites," in *Mechanics of Solids, The Rodney Hill Anniversary Volume,* edited by H. G. Hopkins and M. J. Sewell, pp. 653–686, Pergamon Press, Oxford, 1982). The reader is referred to these sources for further information on how P may be calculated.

A Simple Nonlinear Example

The solution of (21.34), in the case of nonlinear material behavior, generally requires numerical computation. A simple case for which this is not so is provided by a nonlinear matrix $(r=1)$ reinforced by rigid inclusions $(r=2)$. Then for the inclusions,

$$\begin{aligned} W_2(e) &= 0, \qquad e=0 \\ &= +\infty \qquad \text{otherwise.} \end{aligned} \tag{21.39}$$

Correspondingly,

$$(W_2 - W_0)^*(\tau) = 0 \quad \text{identically, so } \gamma_2 = 0. \tag{21.40}$$

(21.34) imply that

$$\sum c_r \gamma_r = \bar{e},$$

and hence in this case that

$$\gamma_1 = \frac{\bar{e}}{c_1}. \tag{21.41}$$

Then, assuming differentiability of $(W_1 - W_0)^{**}$,

$$\tau_1 = (W_1 - W_0)^{**\prime}\left(\frac{\bar{e}}{c_1}\right). \tag{21.42}$$

The polarization τ_2 is now obtained by reverting to (21.34) with $r=2$. This gives

$$\tau_2 = \tau_1 + P^{-1}\left(\frac{\bar{e}}{c_1}\right). \tag{21.43}$$

Substitution into the bound formula (21.39) yields

$$\tilde{W}(\bar{e}) \geq \frac{1}{2}\bar{e}\left(L_0 + \frac{c_2}{c_1}P^{-1}\right)\bar{e} + c_1(W_1 - W_0)^{**}\left(\frac{\bar{e}}{c_1}\right). \tag{21.44}$$

Thus, the bound is completely explicit, in this case. It applies for any choice of L_0, and is nontrivial as long as W_1 grows at least quadratically when $\|e\|$ is large. The best bound is obtained by maximizing (21.44) with respect to L_0; this depends, of course, on the form of W_1.

Further details are not presented, except for the case of a linear matrix characterized by a tensor of moduli L_1. It is necessary to choose L_0 so that $W_1 - W_0$ is convex, and so equal to $(W_1 - W_0)^{**}$. The bound that is obtained thereby also follows as the limiting form of (21.38), as $L_2 \to \infty$.. The optimal bound is obtained, in fact, by choosing $L_0 = L_1$. The resulting bound is

$$\tilde{W}(\bar{e}) \geq \frac{1}{2}\bar{e}\left(L_1 + \frac{c_2}{c_1}P^{-1}\right)\bar{e}. \tag{21.45}$$

This is well known in the special case of isotropic statistics (when P takes a particular form) but not at the level of generality allowed here.

Exploitation of the bound formula (21.39)—and variants—is still in progress at the time of writing. The most detailed computations so far published are those by R. Dendievel, G. Bonnet, and J. R. Willis ("Bounds for the creep behavior of polycrystalline materials," in *Inelastic Deformation of Composite Materials,* edited by G. J. Dvorak, pp. 175–192, Springer-Verlag, New York, 1991).

References

N. I. Akhezier, *The Calculus of Variations*, Blaisdell Publ. Co. (1961).

A. I. Beltzer, *Variational and Finite Element Methods*, Springer-Verlag (1990).

G. A. Bliss, *Lectures on the Calculus of Variations*, Phoenix Science Series, The Univ. of Chicago Press (1946).

O. Bolsa, *Vorlesungen uber variations rechnung*, Chesea Publ. Co. (1909).

C. A. Brebbia, J. C. F. Telles, and L. C. Wrobel, *Boundary Element Techniques*, Springer-Verlag, NewYork (1984).

C. Caratheodary, *Calculus of Variations and Partial Differential Equations of the First Order*, Holdenday, Inc. (1967).

R. W. Cottle, F. Giannessi, and J. L. Lions, eds., *Variational Inequalities and Complementary Problems*, John Wiley & Sons (1980).

R. Courant and D. Hilbert, *Methods of Mathematical Physics*, vol 1, Interscience Pub., Inc. (1953).

J. W. Craggs, *Calculus of Variations*, Crane, Russak & Co. (1973).

L. M. Delves and J. L. Mohamed, *Computational Methods for Integral Equations*, Cambridge Univ. Press (1985).

I. Ekeland and R. Teman, *Convex Analysis and Variational Problems*, vol 1, North Holland (1976).

G. M. Ewing, *Calculus of Variations with Applications*, W. W. Norton and Co. (1969).

G. Fox, *An Introduction to the Calculus of Variations*, Oxford (1950).

P. Funk, *Variations rechnung und ihre anwendung in physik und technik*, Springer-Verlag (1962).

F. D. Gakhov, *Boundary Value Problems*, Dover (1990).

I. M. Gelfand and S. V. Fomin, *Calculus of Variations*, Prentice-Hall, Inc. (1963).

H. H. Goldstine, *A History of the Calculus of Variations from 17th through the 19th Century*, Springer-Verlag (1980).

T. W. Gray and J. Glynn, *Exploring Mathematics with Mathematica*, Addison-Wesley (1991).

P. A. Griffiths, *Exteriror Differential Systems and the Calculus of Variations*, Birkhauser (1983).

R. Herman, *Differential Geometry and the Calculus of Variations*, Academic Press (1968).

M. R. Hestenes, *Calculus Of Variations and Optimal Control Theory*, John Wiley & Sons (1966).

A. D. Ioffe and V. M. Tihomirov, *Theory of External Problems*, NorthHolland (1979).

L. W. Kantorowitsch and W. I. Krylow, *Naherungsmethoden der hoher analysis, Verlag der Wissenschaften*, Berlin (1956).

C. Lanczos, *The Variational Principles of Mechanics*, University of Toronto Press, Tronto (1949).

H. L. Langhaar, *Energy Method in Applied Mechanics*, John Wiley & Sons, Inc., New York (1962).

G. Leitmann, *The Caluculus of Variations and Optimal Control*, Plenum Press (1981).

J. D. Logan, *Invariant Variational Principles*, Academic Press (1977).

J. Mathews and R. L. Walker, *Mathematical Methods of Physics*, Addison-Wesley (1970).

S. G. Mikhlin, *Variational Methods in Mathematical Physics*, Pergamon Press, New York (1964).

H. W. Milnes, *Calculus of Variations*, Texas Technological College, Lubbock (1969).

C. B. Morrey, *Multiple Integrals in the Calculus of Variations*, Springer-Verlag (1966).

R. Maeder, *Programming in Mathematica*, 2nd ed., Addison-Wesley (1991).

M. Morse, *Variational Analysis, Critical Extremals and Sturmian Extensions*, John Wiley (1973).

F. D. Murnaghan, *The Calculus of Variations*, Spartan Books (1962).

S. Nemat-Nasser, (ed.), *Variational Methods in the Mechanics of Solids*, Proceedings of the IUTAM Symposium on Variational Methods in the Mechanics of Solids held at Northwestern University, Pergamon Press, Evanston (1980).

L. A. Pars, *An Introduction to the Calculus of Variations*, Heinemann (1963).

K. Rektorys, *Variational Methods in Mathematics, Science, and Engineering*, D. Reidel Publ. Co. (1977).

H. Rund, *The Hamilton-Jacobi Theory in the Calculus of Variations*, D. Van Nostrand Co. (1966).

D. L. Russel, (ed.), *Calculus of Variations and Control Theory*, Academic Press (1976).

D. R. Smith, *Variational Methods in Optimization*, Prentice-Hall (1974).

V. M. Tikhomirov, *Fundamental Principles of the Theory of External Problems*, John Wiley & Sons (1982).

K. Washizu, *Variational Methods in Elasticity and Plasticity*, Pergamon Press (1968).

K. Washizu, *Variational Methods in Elasticity and Plasticity*, Pergamon Press (1975).

R. Weinstock, *Calculus of Variations, with Applications to Physics and Engineering*, McGraw Hill (1952).

R. Weinstock, *Calculus of Variations*, McGraw Hill Book Co., Inc., New York (1952).

S. Wolfram, *Mathematica: A System for Doing Mathematics by Computer*, 2nd ed., Addison-Wesley (1991).

Appendix

Mathematica Program Listings

Mathematica, published by Wolfram Research, Inc., offered a new approach to science, engineering, and mathematics. Mathematica is a symbolic manipulation code (SMC) which has been around for some time. MACSYMA, for example, is one of the oldest and most popular program for the mainframe and workstation users. Mathematica is one of the first SMCs to become available for personal computers. SMCs which have recently become more readily available to the public, run on various platforms including Apple Macintoshes, IBM PC's, NeXT computers, Sun and other popular workstations. All versions have the same functionalities that no program modification is necessary to port programs from one platform to the other.

Mathematica like most SMCs can manipulate and evaluate mathematical equations both symbolically and numerically. It can also display them graphically. Since most operations are done nearly automatically, users can concentrate on verifying their ideas without being distracted by tedious manipulation details. For this reason, SMCs are educationally very advantageous because instructors can have their students experiment with new theories and methods without the students losing interest because of lengthy and tedious details. Furthermore, instructors can encourage students to try many "what if" cases. The ambitious students may be able to find fine details of the theories and methods or discover correlations among their previously learned theories and methods.

As for the down side, SMCs such as Mathematica require a huge amount of memory. In the case of the Macintosh version (v2.0), 5 megabytes are required; for the Sun workstation version, 7 megabytes. However, more memory is strongly recommended. The performance improves significantly as the size of available memory increases. Although Mathematica supports virtual memory, adding more physical memory (random access memory, RAM) is recommended for the Macintosh and PC versions. The programs listed here should run in the minimum environment. The version used here is Mathematica v2.0 running under System 7 on a Macintosh IIfx.

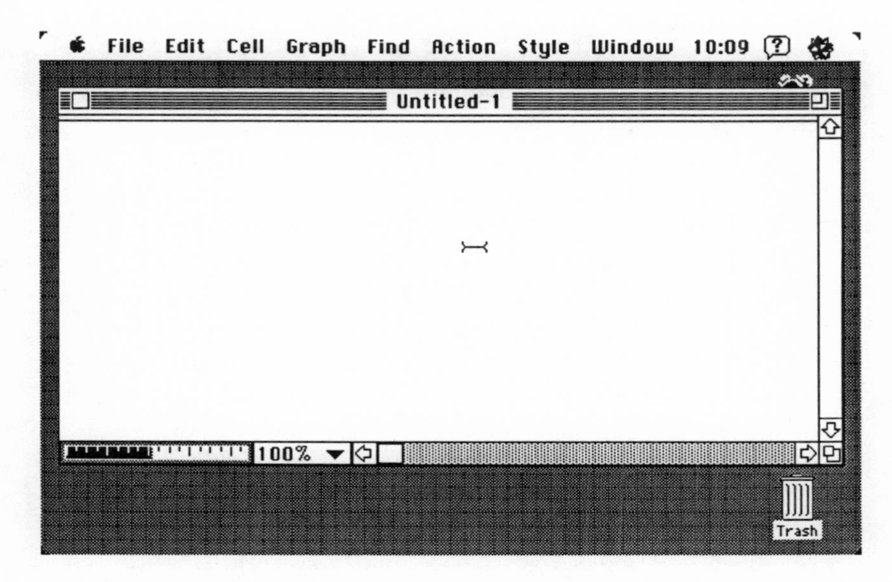

Figure A1. Mathematica display on the Macintosh personal computer.

Using Mathematica
General

Fig. A1 shows a Mathematica display on the Macintosh computer running under System 7. In the white blank area all calculations are performed. The words on top of the screen are called *menus* whence the user can access Mathematica commands. The ticked half black and half white strip on the lower left hand side is the memory indicator displaying how much memory is left for computation. In this case, the black strip indicates about 2 megabytes are already used, and the white blank area indicates about 2.5 megabytes are still free for calculation. The number (100%) next to the memory indicator shows at what magnification the user is viewing Mathematica—in this case, at 100% or the normal magnification.

A Mathematica document file is called a *notebook*. Just as with a paper notebook, the user can jot down ideas, perform mathematical calculations, and plot results. Although Mathematica does not have a drawing environment, the user can create figures in other drawing applications and "paste" electronically into the notebook.

Let us see how Mathematica performs calculations. From the keyboard, type 1+1 and press the <enter> key—not the <return>—immediately after the second 1. The space between the

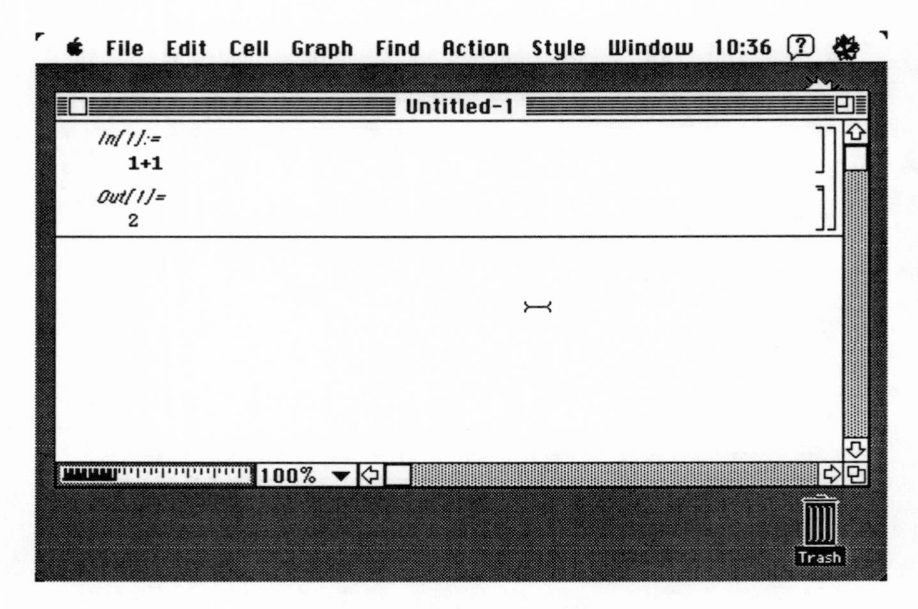

Figure A2. Mathematica calculation.

number and the plus sign is immaterial. The display then appears as shown in Fig. A2. If this command is the very first one after Mathematica has been loaded and executed, Mathematica spends some time loading the core part of the program called the *kernel* into memory. During this time, the 1+1 operation is not performed. The operation is performed only after the kernel has been loaded. The *In[1]* means that the operation is the very first input since Mathematica has been executed, and the *Out[1]* means that the output is the very first evaluation by Mathematica since the program has been executed.

Likewise, the subtraction is done by

```
In[2]:=
    2-7

Out[2]=
    -5
```

The multiplication is done by either

```
In[3]:=
    4*5

Out[3]=
    20
```

or

```
In[4]:=
   4 5

Out[4]=
   20
```

and the division is done by

```
In[5]:=
   1/3

Out[5]=
   1
   -
   3
```

Notice that the division does not yield the value 0.33333... because the fraction cannot be represented in finite precision. We can obtain the numerical value by

```
In[6]:=
   N[%]

Out[6]=
   0.333333
```

where the percentage sign refers to the immediately preceding output which is *Out[5].*

The values of special constants are not expressed numerically in Mathematica, either. For example, the value of π is expressed as **Pi** in Mathematica. But we can always obtain the numerical value by

```
In[7]:=
   N[Pi,20]

Out[7]=
   3.1415926535897932385
```

The second argument 20 is the number of digits to be used to express the first argument **Pi**.

Mathematica also allows for the execution of multiple lines of command at the same time. For example,

```
In[9]:=
   Print["Hello"]
   Print["Good-bye"]

   Hello
   Good-bye
```

where the second line has been entered after pressing the <return> key at the end of the first line.

Sometimes, we do not wish Mathematica to show the calculation result because, for instance, it is an intermediate result and not the final one. To suppress Mathematica from showing the result, we can add a semicolon at the end of the line. For example,

```
In[10]:=
   angle = Pi / 3;
```

Mathematica does not show the result. Line 10 is assigning $\pi/3$ to a variable named *angle*. We can now manipulate this variable.

```
In[11]:=
   Sin[angle]

Out[11]=
   Sqrt[3]
   -------
      2
```

Calculus

Mathematica is an ideal tool for calculus. It can differentiate, integrate, expand in series, solve ordinary differential equations, etc. For instance,

```
In[12]:=
   D[x^2 + Exp[3x] - Sin[x],x]

Out[12]=
       3 x
    3 E     + 2 x - Cos[x]


In[13]:=
   Integrate[x/(a x + b),x]
```

Out[13]=

```
x   b Log[b + a x]
- - --------------
a         2
         a
```

In[14]:=

```
Series[Cos[x],{x,0,6}]
```

Out[14]=

```
      2    4    6
     x    x    x          7
1 - -- + -- - --- + O[x]
     2   24   720
```

In[15]:=

```
DSolve[x^2 u''[x]+x D[u[x],x]-
        k u[x]==0,u[x],x]
```

Out[15]=

```
              C[1]        Sqrt[k]
{{u[x] -> -------- + x          C[2]}}
             Sqrt[k]
            x
```

In[15] has been broken into two lines by pressing the <return> key after the minus sign at the end of the first line.

Linear Algebra

Matrices and vectors are called *lists* in Mathematica. For example, the vector $\mathbf{v} = \{3, 4, 5\}^T$ is represented as

In[16]:=

```
v = {3,4,5}
```

Out[16]=

```
{3, 4, 5}
```

A 2×2 matrix is represented as

In[17]:=

```
m = {{a,b},{c,d}}
```

```
Out[17[=
   {{a, b}, {c, d}}
```

This matrix can be displayed in the matrix form as

```
In[18]:=
   MatrixForm[%]

Out[18]=
   a   b
   c   d
```

Lists can also be generated by the following two commands:

```
In[19]:=
   Table[Sin[n Pi/6],{n,0,6}]

Out[19]=
        1  Sqrt[3]      Sqrt[3]  1
   {0, -, -------, 1, -------, -, 0}
        2     2            2     2
```

or

```
In[20]:=
   consts = Array[c,5]

Out[20]=
   {c[1], c[2], c[3], c[4], c[5]}
```

Mathematica can also solve a system of simultaneous equations analytically

```
In[21]:=
   Solve[{x + 3y + z ==2,
          -x - 2y + z ==5,
          3x + 7y + z ==-3},
          {x,y,z}]

Out[21]=
              13                      5
   {{x -> -(--), y -> 2, z -> -}}
              2                       2
```

Graphics

The graphics capabilities of Mathematica help the user to visualize ideas. In addition, Mathematica supports various 2- and 3-D graphics. Here are some examples:

In[22]:=
```
Plot[Sin[x],{x,0,2Pi}]
```

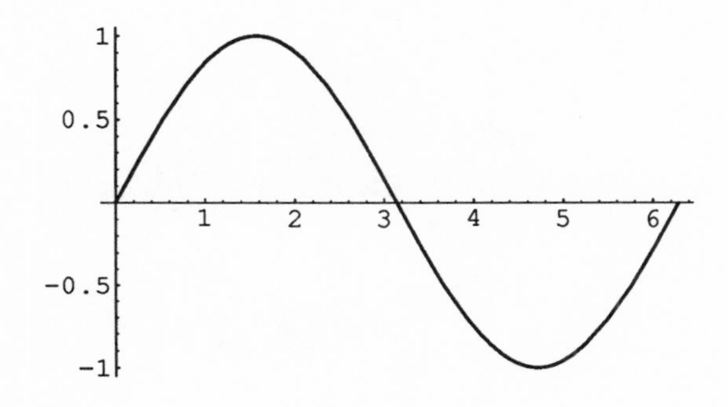

Out[22]=
```
-Graphics-
```

In[23]:=
```
Plot3D[Sin[x]Cos[y],{x,-2Pi,2Pi},
        {y,-2Pi,Pi}]
```

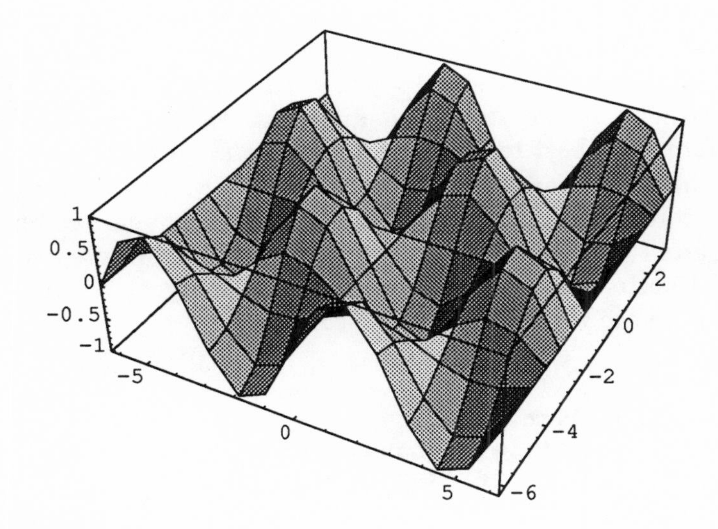

Out[23]=
```
-SurfaceGraphics-
```

In[24]:=
```
u = Sin[rho] (3 + Cos[theta]);
v = Cos[rho] (3 + Sin[theta]);
z = Sin[theta];
ParametricPlot3D[{u,v,z},
  {rho,0,2Pi},{theta,0,2Pi}]
```

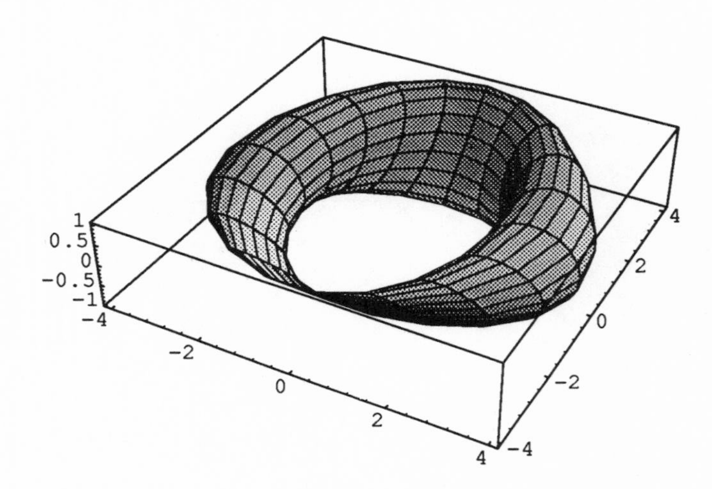

Programming Mathematica

The previous sections have introduced only a few Mathematica commands. Readers are encouraged to consult the Mathematica manual for more commands. If readers are continuing from the previous section, they should quit Mathematica once and restart it. In this section, we explore a different way to use Mathematica; therefore, it is best to restart from scratch.

Although there are nearly 850 built-in commands for Mathematica, its true power is that the user can combine these commands to create new commands. For example, suppose we would like to compute the sum of all odd integers between 1 and n. We can create the following new command:

In[1]:=
```
OddSum[n_] :=
  Module[{i,total=0},
    Do[If[OddQ[i],total+=i,0],
      {i,n}];
    total
    ]
```

Mathematica does not return anything. `OddSum[]` is the name of the command whose argument is the dummy variable `n` which means that the name of this variable is immaterial. The `Module[]` block is the main body of the command. The variable `i` and `total` are called *local* variables that do not have any meaning outside the `Module` block, and `total` is initially set to zero. The `Do[]` block is a loop whose index is `i` and runs from 1 to `n`. `OddQ[]` checks whether its argument `i` is odd, and if it is, `total` is incremented by `i`. The final value is returned in the end.

Let us try this new command. Type

```
In[2]:=
   OddSum[100]

Out[2]=
   2500
```

The command works as it should. This procedure, however, is shown here just to illustrate Mathematica's programmable feature. Usually there is more than one way to write a procedure that does the same task. For instance, we could have written the above procedure the following way:

```
In[3]:=
   OddSum[n_]:=
     Module[{},
       If[OddQ[n],++n,0];
       n^2/4
     ]
```

This procedure is far more efficient than the original one. The reader is encouraged to rewrite the procedures in this appendix in different ways to learn more about Mathematica programming.

We can write more complex and flexible commands. The advantage of programming Mathematica is that many repetitious procedures can be automated, so that users can test different cases as many times as they wish. In the remaining sections, we explore the use of Mathematica for the variational methods.

Variational Derivatives and Euler Equations

We are going to begin by simulating the variational differentiation operation and extracting the Euler equation as described in Chapter 2. But before we start, let us restart Mathematica and turn off a few Mathematica warning features

In[1]:=
```
Off[General::spell1,General::spell1,
  General::intinit]
```

Now, let us take up a very simple functional $I[u]$

In[2]:=
```
F[y_] := 1/2 D[y,x]^2 - 10 y
```

This is the integrand of the functional for the string problem (2.39). Let the domain be $[a,b]$.

We need to define the trial function (2.3)

In[3]:=
```
trialFunc = y[x] + alpha zeta[x]
```

Out[3]=
```
y[x] + alpha zeta[x]
```

Now, we need to substitute this **trialFunc** into the functional **F** and evaluate its derivative with respect to **alpha** at **alpha** = 0

In[4]:=
```
D[F[trialFunc],alpha]/.alpha->0
```

Out[4]=
```
-10 zeta[x] + y'[x] zeta'[x]
```

The **/.** symbol means to apply the right-hand side of the symbol, i.e., **alpha->0** which means to set the value of **alpha** to be zero, to the left-hand side of the symbol which is **D[F[trialFunc],alpha]**.

Out[4] has the form similar to the integrand of the second term on the right-hand side of (2.23). Therefore, the integration of *Out[4]* is the first variation δI. Before we integrate, since we also know that **zeta[x]** vanishes at the end points because the string is fixed at both ends, let us add these conditions at the same time

In[5]:=
```
Integrate[%,{x,a,b}]/.{zeta[a]->0,
  zeta[b]->0}
```

Out[5]=
```
-Integrate[zeta[x] (10 + y''[x]),
   {x, a, b}]
```

In order for the first variation to vanish for any arbitrary **zeta[x]**, the coefficient of **zeta[x]** in the integrand must vanish, and the coefficient is the Euler equation

In[6]:=
```
Coefficient[%[[2]][[1]],zeta[x]]
```

Out[6]=
```
10 + y''[x]
```

which agrees with (2.40) when $T = 1$ and $f = 10$. Perhaps *In[6]* needs some explanations. The **Coefficient[f,c]** command means to collect the coefficient of **c** in the expression **f** where **c** in this case is **zeta[x]**. The expression **%[[2]][[1]]** means to extract the first part of the second part of *Out[5]* in this context. *Out[5]* is composed of 2 parts: The first part is −1, and the second part is **Integrate [zeta[x] (10 + y''[x]), {x, a, b}]**. In turn, the first part of the second part of *Out[5]* is **zeta[x] (10 + y''[x]),** and the second part is **{x, a, b}**. Therefore, the coefficient of **zeta[x]** in **%[[2]][[1]]** is indeed *Out[6]*.

Finally, we can solve *Out[6]* for **y[x]** with the corresponding boundary conditions **y[a] = 0** and **y[b] = 0**

In[7]:=
```
DSolve[{%==0,y[a]==0,y[b]==0},y[x],x]
```

Out[7]=
```
{{y[x] -> 5 (a - x) (-b + x)}}
```

Consequently, the final solution is

In[8]:=
```
y[x] = y[x]/.(%[[1]][[1]])
```

Out[8]=
```
5 (a - x) (-b + x)
```

Let us plot this result assuming **a = 0** and **b = 1**

In[9]:=
```
Plot[(%16/.{a->0,b->1}),{x,0,1},
  AxesLabel->{x,y}]
```

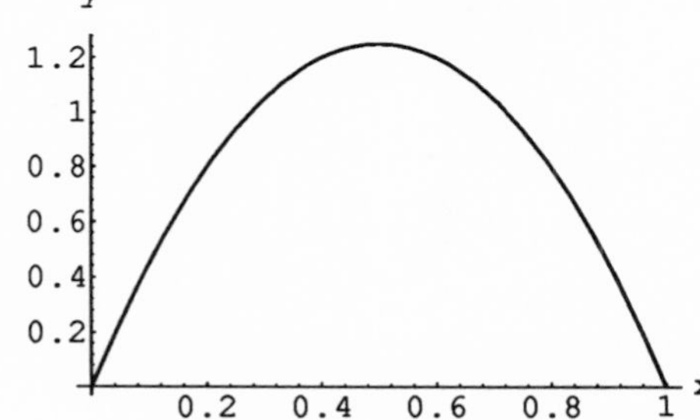

```
Out[9]=
   -Graphics-
```

The purpose of this section is to show the mechanism of the variational differentiation and the procedure for obtaining the Euler equation. Notice that the user does not have to worry about the manipulation details. The required commands nearly follow the main body of the procedure without sidetracking into smaller details.

Ritz's Method

Ritz's method introduced in Chapter 3 is easy to use when dealing with simple functionals and trial functions. For more complex problems, the manual execution of the method is nothing but tedious. With the help of Mathematica, not only can the method be used fairly easily for complicated problems, but also it can assist the instructor to teach the basic ideas of this method.

Let us begin afresh. As before, let us turn off the same warning features of Mathematica by the `Off[]` command

```
In[1]:=
   Off[General::spell1,General::spell1,
       General::intinit]
```

Let us solve the cantilever beam problem. First, we need the functional (4.6) over the domain [0,1]

```
In[2]:=
   F[u_] := Integrate[1/2 D[u,{x,2}]^2 -
                10 u,{x,0,1}]
```

for which we assume $EI = 1$ and $f = 10$.

Next, we need a trial function `y[x]` that satisfies the boundary conditions `y[0] = 0` and `y'[0] = 0`. We choose a fourth order polynomial function, and it needs 5 constants

```
In[3]:=
   constants = Array[c,5]
```

```
Out[3]:=
   {c[1], c[2], c[3], c[4], c[5]}
```

Here is a fourth order polynomial function

```
In[4]:=
   trialFunc = Sum[c[i] x^(i-1),{i,5}]
```

Out[4]:=

```
                        2           3
c[1] + x c[2] + x  c[3] + x  c[4] +

   4
  x  c[5]
```

This function still cannot be the trial function because it does not satisfy the boundary conditions. We let Mathematica determine the trial function

In[5]:=

```
trialFunc = trialFunc/.
               Solve[
                 {(trialFunc/.{x->0})==0,
               (D[trialFunc,x]/.{x->0})==0},
                      constants][[1]]
```

Out[5]=

```
 2           3           4
x  c[3] + x  c[4] + x  c[5]
```

Let us check the boundary conditions again

In[7]:=

```
trialFunc/.x->0
D[trialFunc]/.x->0
```

Out[6]=

0

Out[7]=

0

Therefore, **trialFunc** is an admissible function. Now, the next step is to substitute **trialFunc** into **F[]** .

In[8]:=

```
F[trialFunc]
```

Out[8]=

```
       2
2 c[3]  + 6 c[3] c[4] +

  c[4] (-5 + 36 c[5])
  ------------------- +
           2
```

```
2 c[5] (-5 + 36 c[5])
--------------------- +
          5

                               2
2 (-5 c[3] + 9 c[4]  + 12 c[3] c[5])
-------------------------------------
                  3
```

We have an ordinary function of the constants **c[3]**, **c[4]**, and **c[5]** We calculate the gradient of *Out[8]*

In[9]:=

```
gradient = Table[D[%,c[i]],{i,5}]
```

Out[9]=

```
{0, 0, 4 c[3] + 6 c[4] +

  2 (-5 + 12 c[5])
  ----------------,
         3

                         -5 + 36 c[5]
 6 c[3] + 12 c[4] + ------------,
                             2

                         72 c[5]
 8 c[3] + 18 c[4] + ------- +
                            5

  2 (-5 + 36 c[5])
  ----------------}
         5
```

The components of this gradient must vanish individually

In[9]:=

```
Solve[gradient==0,constants][[1]]
```

Out[9]:=

```
             5              5              5
{c[3] -> -, c[4] -> -(-), c[5] -> --}
             2              3             12
```

Therefore, the final solution is

In[10]:=

```
ritz = trialFunc/.%
```

Out[10]=

$$\frac{5\ x^{2}}{2} - \frac{5\ x^{3}}{3} + \frac{5\ x^{4}}{12}$$

and Ritz's method is done. We can plot this solution

In[11]:=
```
Plot[ritz,{x,0,1},
  AxesLabel->{"x","y"}]
```

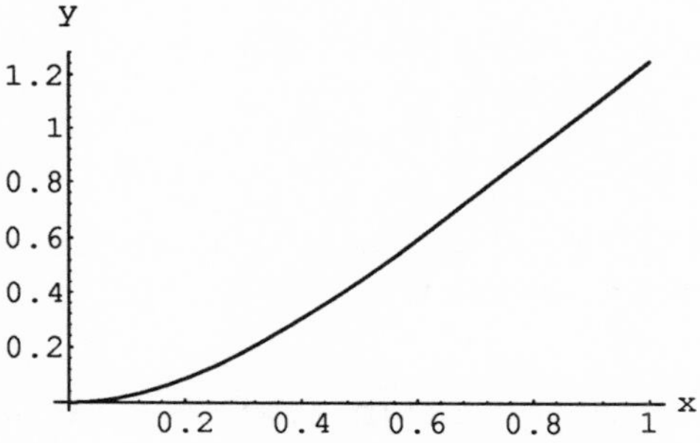

Out[11]=
```
-Graphics-
```

The user can try different functionals and trial functions to study the behavior and the mechanism of this method.

Programming Ritz's Method

The procedure in the previous section can be automated. **Ritz1D** listed below does just that. The user needs to supply the functional, the trial function that satisfies the boundary conditions, the constants in the trial function, and the domain. **Ritz1D** returns the solution by Ritz's method. Here is the listing:

In[12]:=
```
Ritz1D::usage =
"Ritz1D[Fn,TrialFunc,Consts] applies
Ritz's method to the functional Fn using
the trial function TrialFunc composed of
the constants Consts.";
```

In[13]:=
```
Ritz1D[Fn_, TrialFunc_, Consts_]:=
  Module[{F,grad,i,n=Length[Consts]}
    F = Fn[TrialFunc];
    grad = Table[D[F,Consts[[i]]],{i,n}];
    TrialFunc/.Solve[grad==0,Consts][[1]]
  ]
```

In[12] is the description of **Ritz1D** which can be called up by

In[14]:=
```
?Ritz1D
```

```
Ritz1D[Fn,TrialFunc,Consts] applies
   Ritz's method to the functional Fn
   using the trial function TrialFunc
   composed of the constants Consts.
```

Let us use **Ritz1D**. First, we need to define the functional, the constants, and the trial functions

In[15]:=
```
g[u_]:= Integrate[(D[u,x]^2-(1+x^2)u^2-
                        2u)/2,{x,-1,1}];
consts = {c1,c2};
u = c1(1-x^2)+c2(1-x^4);
```

In[16]:=
```
Ritz1D[g,u,{c1,c2}]
```

Out[16]=

$$\frac{1050\,(1 - x^2)}{1063} - \frac{231\,(1 - x^4)}{4252}$$

The advantage of using a procedure is that once the mechanism of the method is understood, the user can experiment with different problems with various trial functions thereby strengthening the understanding of the method. **Ritz1D** can be improved to automate even further; for instance, the user can modify the procedure in such a way that Mathematica automatically chooses the proper trial function based on the user-supplied boundary conditions and continuity conditions.

Galerkin's Method

Let us now use Mathematica to perform Galerkin's method. We examine the same cantilever problem solved in the previous section. As before, we restart Mathematica and turn off a few warnings.

```
In[1]:=
   Off[General::spell,General::spell1,
       General::intinit]
```

We define the Euler equation (4.8)

```
In[2]:=
   dF[u_] := D[u,{x,4}] - 10
```

Next, we use the same fourth order polynomial function

```
In[3]:=
   consts = Array[c,5];
   trialFunc = Sum[c[i] x^(i-1),{i,5}];
```

This polynomial function must satisfy not only the essential boundary conditions but also the natural boundary conditions

```
In[4]:=
   trialFunc =
     trialFunc/.
       Solve[{(trialFunc/.x->0)==0,
              (D[trialFunc,x]/.x->0)==0,
              (D[trialFunc,{x,2}]/.x->1)==0,
              (D[trialFunc,{x,3}]/.x->1)==0}
       ][[1]]

Out[4]=

                   3           4
  2             2 x  c[3]     x  c[3]
 x  c[3] - --------- + -------
                 3             6
```

The weight function is just the derivative of **trialFunc** with respect to **c[3]**

In[5]:=

```
weight = D[trialFunc,c[3]]
```

Out[5]=

```
         3    4
 2    2 x    x
x  -  ---- + --
       3     6
```

We are now in the position to evaluate (8.11)

In[6]:=

```
Integrate[dF[trialFunc] weight,{x,0,1}]
```

Out[6]=

```
5 - 2 c[3]   11 (-5 + 2 c[3])
---------- + ----------------
    3               15
```

Now, all we have to do is solve for **c[3]** and substitute it back into **trialFunc**

In[7]:=

```
galerkin = trialFunc/.
                Solve[%==0,c[3]][[1]]
```

Out[7]=

```
   2      3      4
5 x    5 x    5 x
---- - ---- + ----
 2      3      12
```

As expected, the solution is the same as the one in the previous section. However, we must understand that the reason we obtain the same result for both methods is that the form of the trial function happens to be that of the exact solution. In general, these methods yield slightly different solutions. Users should try various problems.

Basically, the operations of Ritz's and Galerkin's methods are not so different. Once the user understands the Ritz operation, the Galerkin operation can be mastered very quickly. The above operations are used, as in the previous section, to learn the mechanism of Galerkin's method. Users should try different problems several times until they feel comfortable with the operations.

Programming Galerkin's Method

After users have familiarized themselves with the mechanism of Galerkin's method, they can automate the procedure and explore the characteristics of the method. The procedure listed below is a simple implementation of Galerkin's method in Mathematica

```
In[8]:=
   Galerkin1D::usage =
   "Galerkin1D[dF,TrialFunc,
   Consts,{x,x0,x1}] applies Galerkin's
   method to the differential equation dF
   using the trial function TrialFunc
   composed of the constants Consts over the
   domain of x ranging from x0 to x1.";

In[9]:=
   Galerkin1D[dF_, TrialFunc_,
     Consts_,{x_,x0_,x1_}] :=
       Module[{gErr,n=Length[Consts],
               totalErr},
         gErr = Table[dF[TrialFunc] *
                   D[TrialFunc,Consts[[i]]],
                      {i,n}];
         totalErr =
           Integrate[gErr,{x,x0,x1}];
           TrialFunc/.
             Solve[totalErr==0,Consts][[1]]
   ]
```

Let us see how this procedure works

```
In[11]:=
   g[u_] := D[u,{x,2}] + (1+x^2) u + 1;
   h = c1(1-x^2) + c2(1-x^4);

In[12]:=
   Galerkin1D[g,h,{c1,c2},{x,-1,1}]

Out[12]=

                  2                 4
   1050 (1 - x )     231 (1 - x )
   --------------- - --------------
        1063              4252
```

This problem corresponds to the problem solved by `Ritz1D`. The user is encouraged to try different trial functions and improve this procedure.

Eigenvalues
We now take a look at eigenvalue problems. Example 16.1 is examined here. (16.12) and (16.13) are

In[1]:=
```
Off[General::spell1,General::spell,
    General::intinit]
```

In[4]:=
```
F[u_]:= Integrate[D[u,x]^2/2-(1+w)u^2/2,
           {x,0,1}];
Consts = {a1,a2};
u = x(x-1)(a1 + a2 x);
```

We substitute **u** into **F** and extremize with the constants **a1** and **a2**

In[5]:=
```
Table[D[F[u],Consts[[i]]],{i,2}]
```

Out[5]=
```
                   (a1 - a2) (1 + w)
{-a1 + a2 + ----------------- -
                          4

   a2 (1 + w)
   ---------- +
       6

   6 a1 - 14 a2 - 2 a1 w
   --------------------- +
             6

   a1 + 6 a2 + a1 w
   ---------------- +
          4

   -2 a1 + 4 a2 - 2 a1 w + 4 a2 w
   ------------------------------,
                 10

       3 (a1 - a2)    -14 a1 + 8 a2
  a1 + ----------- + ------------- +
            2              6

   a2 (1 + w)    (-a1 + a2) (1 + w)
   ---------- + ------------------ -
       42               6
```

```
  a1 + 6 a2 + a1 w
  ---------------- +
         4

  4 a1 + 16 a2 + 4 a1 w - 2 a2 w
  ------------------------------}
                10
```

We can simplify *Out[5]*

In[6]:=
```
Simplify[%]
```

Out[6]=
```
  (2 a1 + a2) (9 - w)
{-------------------,
          60

  63 a1 + 52 a2 - 7 a1 w - 4 a2 w
  -------------------------------}
                420
```

Now, we need to convert *Out[6]* into the matrix form (16.15). The coefficient matrix of (16.15) is

In[7]:=
```
Table[Coefficient[ExpandAll[%[[i]]],
      Consts[[j]]],{i,2},{j,2}]
```

Out[7]=
```
   3    w    3    w
{{-- - --, -- - --},
  10   30   20   60

   3    w   13     w
 {-- - --, --- - ---}}
  20   60  105   105
```

The roots of the determinant of *Out[7]* are the approximations of the eigenvalues

In[8]:=
```
Solve[Det[%15]==0,w]
```

Out[8]=
```
{{w -> 41}, {w -> 9}}
```

which agree with (16.17). We can use higher order polynomials to find more eigenvalues.

Finding Eigenvalues with Mathematica

Finally, here is a procedure to approximate eigenvalues. This procedure is designed for the polynomial approximation, and because there may be no analytical solutions for the roots of higher order polynomials, the **Solve** command has been replaced by the **NSolve** command which solves equations numerically.

```
In[9]:=
  FindEV::usage =
  "FindEV[F,TrialFunc,Consts,Ev]
  numerically approximates the eigenvalues
  Ev of the functional F using the
  polynomial trial function TrialFunc
  composed of the constants Consts.";

In[10]:=
  FindEV[F_, TrialFunc_, Consts_, Ev_] :=
    Module[{i,j,m,EXF,dEXF,
      n=Length[Consts]},
      EXF = ExpandAll[F[TrialFunc]];
      dEXF = Table[D[EXF,Consts[[i]]],
                   {i,n}];
      m = Table[Coefficient[dEXF[[i]],
      Consts[[j]]],{i,n},{j,n}];
      NSolve[Det[m]==0,Ev]
    ]
```

As a check, let us try the same problem but this time using an eighth order polynomial

```
In[11]:=
  consts = Array[c,8];
  u = x(x-1) Sum[c[i] x^(i-1),{i,8)];

In[12]:=
  FindEV[F,u,Consts,w]

Out[12]=
  {{w -> 8.8696}, {w -> 38.4784},
  {w -> 87.8295}, {w -> 156.957},
  {w -> 253.418}, {w -> 375.474},
  {w -> 877.883}, {w -> 1297.09}}
```

Compared with Table 16.1, these values are closer to the exact values as expected.

Index